Microelectronics A to Z

Malcolm Plant

Microelectronics A to Z

Longman

Longman Group Limited,
Longman House, Burnt Mill, Harlow,
Essex CM20 2JE, England
and Associated Companies throughout the world.

First published 1985

British Library Cataloguing in Publication Data
Plant, M. (Malcolm)
  Microelectronics A to Z.
  1. Microelectronics—Dictionaries
  I. Title
  621.381'71'0321   TK7874

  ISBN 0-582-89285-6

Typeset by Computerset (MFK) Ltd.,
Saffron Walden, Essex.

Printed in Great Britain
by Butler & Tanner Ltd.,
Frome & London.

About the author

Malcolm Plant, a physicist, is Principal Lecturer at
Trent Polytechnic, Nottingham.

He has contributed to various Schools' Council
(School Curriculum Development Committee)
projects and 'O' and 'A' level examination syllabuses,
and has researched into space physics. His
publications include *Op Amp Applications* (1974),
Basic Electronics (1975), and *Instrumentation* (1982).

Contents

Introduction

MICROELECTRONICS is about making and using complex circuits on tiny silicon chips. A typical silicon chip has about as many electronic components (eg resistors and transistors) on it as several thousand transistor radios. The microprocessor, a 'computer on a chip', is the best known of the hundreds of thousands of different types of integrated circuits being used in products as close at hand as the digital watch, and as far away as the computers on board Voyager now exploring the outer reaches of our solar system.

Development of miniaturized circuits

For a product with such a short history – the first integrated circuits were made in the 1960s and the first microprocessors in 1972 – it is surprising that silicon chips, and especially the microprocessor, should be making such a big impact on our lives. Why put all the important parts of a computer on a piece of silicon barely distinguishable from a splinter of wood and so small you could hide it under your thumb nail? Of course there is more to a computer than a microprocessor – a keyboard and screen are important too – but what is the point of making circuits so small? The stimulus to do so has come from three main areas: weapons technology; the space race; and commercial activity.

Modern weapon systems depend for their success on circuits which are small, light, quick to respond, and absolutely reliable, and which use hardly any electrical power. Miniature circuits on silicon chips offer all these advantages.

Russia started the space race by successfully launching Sputnik in 1957. At first, the American response to this challenge was unsatisfactory, but by the end of the 1960s the USA was the first nation to land a man on the Moon. Lacking the enormously powerful booster rockets developed by the USSR, the USA needed compact and complex spacecraft. This stimulated the miniaturization of electronic circuits in general and of computers in particular, especially for the complex manoeuvres required of the space shuttle and the remote control of interplanetary probes such as Voyager.

During the 1970s, spin-off from military interests and the space-race stimulated the growth of an electronics industry bent first on creating, then satisfying, the demand for electronics goods in the home, the office, and industry. Since then the consumer market has increasingly dictated its own requirements, especially for microcomputers and other microprocessor-based products.

Making an integrated circuit

The electronic components (eg transistors) and the connections between them which make up the circuits on a silicon chip are the smallest functioning manmade objects ever created and the trend is towards even finer detail. The process of putting several hundred thousand transistors on a silicon chip a few millimetres square is not easy. It begins with a cylinder-shaped single crystal of very pure silicon, known as a 'boule' or 'ingot', which is 'grown' from molten silicon especially for the job of making silicon chips. The ingot is sawn up into thin slices, known as 'wafers', about the size of a beer mat and half as thick.

The wafers are then passed through a chamber containing gases heated to about 1200 degrees celsius. The gases diffuse into each wafer to give it the properties of a 'p-type' or an 'n-type' semiconductor depending on the gas used – a process known as 'epitaxial growth'. The wafers are then ready to have integrated circuits formed within them by a complex process which involves three main techniques: masking, etching, and diffusion. A preliminary masking and etching process produces many thousands of minute 'windows' in the surface of a wafer. Gases are allowed to diffuse through these windows to form components in the underlying silicon. Several steps are necessary to produce a set of transistors, each involving the process of masking and etching as shown in Figure 1.

Masking and etching

The first stage in the process involves heating the wafer to about 1000 degrees celsius in a stream of oxygen so that a thin layer of silicon dioxide is formed over the whole surface of the wafer.

In the next stage, a thin layer of a light-sensitive emulsion (called photoresist) is spread over the layer of silicon dioxide. A photographic plate (a mask) is placed over the top of this emulsion.

The mask contains a pattern of dots in microscopic detail which are to become holes in the silicon dioxide layer. A single mask holds the pattern for several hundred integrated circuits for each wafer.

In the third stage, the mask is exposed to ultraviolet light. Where the mask is transparent, the light passes through and chemically changes the photoresist underneath so that it hardens. The unexposed photoresist can easily be removed with a suitable solvent (fourth stage).

Next, (fifth stage), the silicon wafer is immersed in another solvent which removes the silicon dioxide from the unexposed areas. The wafer now has a thin surface layer of silicon dioxide in which there are a large number of minute 'windows'. It is through these windows that gases are allowed to pass into the silicon layer underneath to form transistors. In the formation of complete silicon chips on a wafer, the formation of a silicon dioxide layer, followed by masking and etching, has to be repeated many times. Figure 2 shows the steps in the process of making a single npn transistor through one of these windows.

Making a transistor

First, a gas is selected which diffuses through a window to form a p-type base region in the n-type silicon layer. Next, a fresh silicon dioxide layer is formed over the window, followed by a stage of masking and etching to create a second smaller window. Through this window, a gas diffuses to form the n-type emitter region. Another layer of silicon dioxide is formed over this window, followed by masking and etching to create smaller windows for making contacts to the base and emitter regions. These contacts are made by depositing aluminium in vapour form.

When all the integrated circuits have been formed on a wafer in this way, the wafer is cut up into individual chips and packaged in a form which can be used by the circuit designer. This is the familiar 'many-legged' device known as an integrated circuit.

Microelectronics today

Over the years this complex process has been so highly developed that hundreds of thousands of integrated circuits are now mass-produced each year at a cost of a few pence each. As a result, many products, and particularly microprocessors, are based on integrated circuits. Since the microprocessor is so small, and it does not need much electrical power to carry out the tasks set by a program of instructions, it can revolutionize the function and reduce the size of many conventional products: electronic portable typewriters and sewing machines, toasters and pacemakers, for example. But the microprocessor has stimulated the creation of many new products: home computers and communications satellites, video disk players, X-ray scanners, and industrial robots are examples of products which would not exist were it not for the invention of the silicon chip.

Steps in the formation of a window in the silicon dioxide surface on a silicon chip

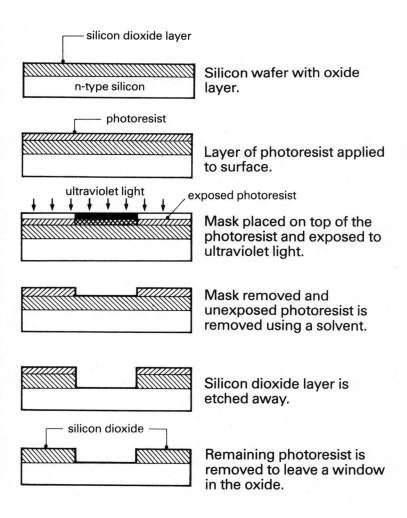

silicon dioxide layer

n-type silicon

Silicon wafer with oxide layer.

photoresist

Layer of photoresist applied to surface.

ultraviolet light / exposed photoresist

Mask placed on top of the photoresist and exposed to ultraviolet light.

Mask removed and unexposed photoresist is removed using a solvent.

Silicon dioxide layer is etched away.

silicon dioxide

Remaining photoresist is removed to leave a window in the oxide.

Figure 1

Steps in the formation of a single NPN transistor on a silicon chip

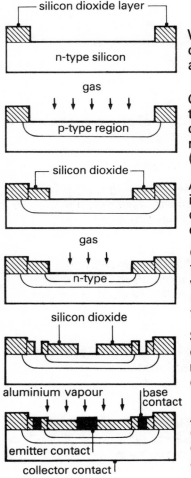

Window created in a silicon dioxide layer by masking and etching.

Gas allowed to diffuse through the window to create a p-type (base) region in the n-type (collector) region.

A smaller window is created in a second layer of silicon dioxide by masking and etching.

Gas is allowed to diffuse through this second window to create an n-type (emitter) region in the p-type region.

Smaller windows created over the n-type and p-type regions by masking and etching.

An aluminium film is deposited, masked and etched to make the emitter and base connections.

Figure 2

Steps in the formation of a single (N-channel) MOSFET transistor on a silicon chip

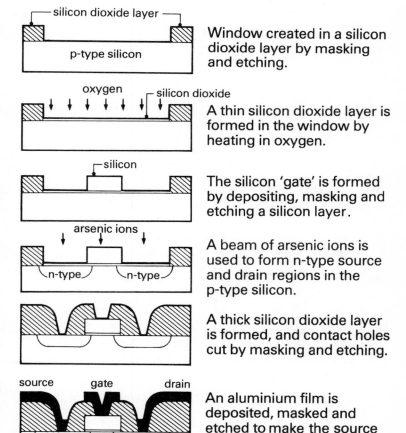

Window created in a silicon dioxide layer by masking and etching.

A thin silicon dioxide layer is formed in the window by heating in oxygen.

The silicon 'gate' is formed by depositing, masking and etching a silicon layer.

A beam of arsenic ions is used to form n-type source and drain regions in the p-type silicon.

A thick silicon dioxide layer is formed, and contact holes cut by masking and etching.

An aluminium film is deposited, masked and etched to make the source and drain connections.

Figure 3

Abbreviations
used in this dictionary

abbr	abbreviation or abbreviated form
adj	adjective
adv	adverb
esp	especially
n	noun
UK	United Kingdom
USA	United States of America
usu	usually
vb	verb

●   SMALL CAPITALS refer the reader to other entries

●   □ a square box followed by 'see' and SMALL CAPITALS refers to other entries where the reader will find related information

●   [ ] square brackets following the part of speech enclose the full form of abbreviated terms

●   ( ) round brackets following the part of speech give information about commonly used abbreviations or acronyms

●   < > angle brackets enclose a phrase illustrating a typical use of the term in context. The term being illustrated is represented by a swung dash ($\sim$).

A

A 1 the symbol for the unit of electrical current, the ampere □ see
AMPERE **2** the symbol used in the hexadecimal system for the decimal
number 10 (eg the hexadecimal number A6 equals the decimal
number 10 × 16 + 6 = 166) □ see HEXADECIMAL

abend *n* [*ab*normal *end*] the unplanned termination of a (computer)
task prior to its completion because of software or hardware errors
which cannot be resolved – compare ABORT □ see CRASH

abort *vb* [*abnor*mal *t*ermination] to deliberately terminate a
(computer) task prior to completion either because of error or to
make way for another job – compare ABEND

absolute addressing *n* DIRECT ADDRESSING

absolute temperature scale *n* a scale of temperature that is
measured from the lowest possible temperature (absolute zero) and
that has the melting point of ice at about 273 degrees kelvin (273K).
Temperatures on the celsius (formerly centigrade) scale are obtained
by subtracting the number 273 from the absolute temperature. Thus
−273 celsius (−273C) is equal to 0K; 288K is equal to 15C, etc. The
absolute temperature scale is the natural scale for all scientific and
engineering studies. □ see CRYOGENIC, CRYOGENIC MEMORY,
JOSEPHSON JUNCTION, SUPERCONDUCTOR

AC – see ALTERNATING CURRENT

acceptor *adj* of or being impurity atoms (eg boron atoms),
introduced into pure silicon to provide a deficiency of electrons –
compare DONOR – **acceptor** *n* □ see P-TYPE, IMPURITY, DIFFUSION

access *vb* to retrieve information from, or to store information in,
locations in a computer memory or storage device <*to ~ a file*>
□ see ACCESS TIME, ACCESS PATH, RANDOM-ACCESS MEMORY

accessory *n* PERIPHERAL

access path *n* a communications link between two or more data
terminals (eg telephones or computer terminals) □ see
COMMUNICATIONS SYSTEM, ACCESS TIME

access time *n* **1** the time interval between a request for information
from a computer memory and the instant this information is
available **2** the time interval between presenting information for
storage in a computer memory device and the instant this
information is stored. Access times are continually being reduced in
order to cope with the demands for computers to process
information at ever faster speeds. □ see MEMORY, GALLIUM
ARSENIDE, JOSEPHSON MEMORY

accordion fold paper *n* FANFOLD PAPER

accumulator *n* **1** one of a number of registers in a microprocessor
for storing binary numbers to be used for arithmetic, logical, and
input/output operations. □ see ARITHMETIC AND LOGIC UNIT **2** a cell
or battery (eg a car battery) that can be recharged after it has

discharged; a secondary cell or connected series of secondary cells

accuracy *n* the nearness of a measured value to its true value <*a digital watch has better ~ than a spring-wound watch*> – **accurate** *adj*, **accurately** *adv*

acoustic *adj* involving sound and its generation, transmission, or use – **acoustics** *n* □ see ACOUSTIC COUPLER, ULTRASONIC SCANNER

acoustic coupler *n* a device connected to a telephone to enable one computer to send information to, and receive information from, other computers. An acoustic coupler has two flexible cups on it into which the telephone handset fits so that there is an efficient transfer of information to and from the computer. □ see MODEM

activate *vb* to get something ready for use <*the burglar alarm was ~d*>

active *adj* of or being an electronic device (eg a transistor) which uses electrical energy to change the amplitude or some other characteristic of a signal – compare PASSIVE

actuator *n* SERVOMECHANISM

Ada *n* a high-level computer language developed mainly for real time programming in scientific, industrial, and military systems; named after Countess Ada Lovelace, a friend of Charles Babbage □ see BABBAGE, REAL TIME PROGRAMMING

adaptive *adj* of or being a machine, esp a computerized machine, that has the ability to adjust to new conditions – **adaptively** *adv*, **adaptivity** *n* □ see ARTIFICIAL INTELLIGENCE, FIFTH GENERATION COMPUTER

ADC – see ANALOGUE-TO-DIGITAL CONVERTER

adder *n* a digital circuit used in calculators and computers which adds together two binary numbers □ see HALF-ADDER, FULL-ADDER

add-on *n* PERIPHERAL

¹address *vb* **1** to identify a storage location in a computer's memory in readiness for a transfer of data between it and another location – compare ACCESS **2** to make an electron beam activate a pixel on the screen of a VDU

²address *n* a number, usu in binary code, that identifies a particular location where data is stored in a computer memory. By means of an address the contents of the cell can be examined or, in the case of random-access memory, altered by a programmer. □ see ADDRESS BUS, MEMORY

address bus *n* an electrical pathway between the microprocessor and memory in a computer that consists of a set of wires (usu tracks on a printed circuit board) along which binary signals flow two bytes at a time that represent a place in the computer's memory where data is stored □ see ADDRESS, ADDRESS BUS, MEMORY

address space *n* the range of addresses used by a computer for storing program instructions □ see ADDRESS, MEMORY MAP

ADSR *abbr* [*a*ttack, *d*ecay, *s*ustain and *r*elease] parts of the amplitude of the envelope of a sound generated by a musical instrument (eg a computer with sound producing facilities). The attack is the immediate rise in the amplitude to its maximum height when a note is first generated. The decay is the settling down period of the note to reach a steady level (the sustain level), and the release is the falling away to silence at the end of the note. □ see ENVELOPE

adventure game *n* a microcomputer program in which the player, as the hero or heroine, has to anticipate and overcome various dangers in the search for treasure in an unknown land. An adventure game is designed as a sort of interactive database which the player uses to reveal critical information to update the course of the adventure. □ see INTERACTIVE

aerial *n* an arrangement of conductors usu placed in an elevated position (eg on or between masts) which transmits and/or receives radio waves <*a television* ~> □ see ANTENNA, DIPOLE AERIAL

AF – see AUDIO FREQUENCY

AFC – see AUTOMATIC FREQUENCY CONTROL

afterglow *n* – see PHOSPHORESCENCE

AGC – see AUTOMATIC GAIN CONTROL

AI – see ARTIFICIAL INTELLIGENCE

air cushion *n* – see WINCHESTER DISK

air traffic control *n* (abbr **ATC**) a computer-assisted system in any busy international airport which provides air traffic controllers with information about the movement of aircraft in their neighbourhood. The position of each aircraft is shown as a blip on a radar screen. Alongside the blip are flags showing the flight number, altitude, and destination of the aircraft. This information is supplemented by computer printouts of the type of aircraft, its flight path, payload, and altitude so that a controller can find the most economical way of guiding the aircraft through his/her area.

Algol *n* [*algo*rithmic *l*anguage] a high-level computer language used mainly for mathematical and scientific applications

algorithm *n* a sequence of steps which make up a plan for solving a particular problem. In computer programming, an algorithm is drawn up as a list of instructions. □ see FLOWCHART, INSTRUCTION

allocate *vb* to earmark complete programs or subroutines to a storage system – **allocation** *n*

allophone *n* an acoustic variant of a phoneme which sounds different depending on its relation to other sounds in a word. For example, the slightly different sounds of 'p' in 'pat' and 'spat' are allophones of the phoneme 'p'. Artificial speech is generated by putting together the 64 or so allophones in the English language using an electronic speech synthesizer. □ see PHONEME, SPEECH SYNTHESIS

alloy *vb* to melt a small quantity of metallic impurity (eg boron) so that it dissolves pure silicon which then hardens to produce a useful semiconductor (eg p-type semiconductor) for making transistors and diodes. This process has largely been superseded by gaseous diffusion – **alloy** *n* □ see IMPURITY, GASEOUS DIFFUSION, PHOTOLITHOGRAPHY

alphabet *n* any character set used to write a (computer) language – **alphabetical** *adj*

alpha-geometrics *n* high-resolution graphics displayed on a VDU using a screen divided into 320 by 240 pixels. As well as the usual letters, numbers, and mosaic characters, alpha-geometrics allows lines, circles, and similar shapes to be drawn using simple commands. Alpha-geometrics are usu provided by a dedicated microcomputer – compare ALPHA-MOSAICS – **alpha-geometric** *adj*

alpha-mosaics *n* low-resolution graphics of alphanumeric characters displayed on a VDU using a screen divided into 40 by 20 squares. These graphics are so-called because pictures are built up by putting together coloured squares as if building up mosaics using tiles. Prestel and other teletext systems use alpha- mosaics. The pictures are low-resolution since only 1 kilobyte of memory is available for each page, and also because higher resolution pictures would take too long to transmit along a telephone line – compare ALPHA-GEOMETRICS – **alpha-mosaic** *adj* □ see TELETEXT, VIEWDATA EDITOR

alphanumeric *adj* displayed or presented as letters, numbers, and some punctuation marks and mathematical symbols <*a hexadecimal number is an ~ code*> – **alphanumerically** *adv*, **alphanumerics** *n*

alternating current *n* (abbr **AC**) an electrical current which reverses its direction at regular intervals <*the mains ~ reverses its direction every one hundredth of a second*> □ see CYCLE, FREQUENCY

ALU – see ARITHMETIC AND LOGIC UNIT

aluminium *n* a metallic element of atomic number 13 which conducts heat and electricity well. It is an important material in the manufacture of integrated circuits. Aluminium is heated in a vacuum so that it evaporates and forms a very thin layer over the entire surface of a silicon chip in the final stages of making an integrated circuit. This thin layer of aluminium is cut into a pattern of conducting paths using the techniques of photolithography and etching. □ see PHOTOLITHOGRAPHY, GOLD, METALLIZATION

AM – see AMPLITUDE MODULATION

ammeter *n* an instrument for measuring the strength of an electric current in amperes or fractions of an ampere (eg milliamperes) □ see AMPERE, MULTIMETER

ampere *n* (symbol **A**) a rate of flow of electrical charge in a circuit equal to 1 coulomb per second □ see COULOMB, OHM'S LAW

ampersand *n* the symbol & used **a** on a microcomputer keyboard to indicate 'and' **b** in a program to indicate hexadecimal numbers **c** in logic diagrams to indicate an AND gate

amplifier *n* a circuit, system, or integrated circuit which increases the electrical potential, current, or power of a signal. Audio amplifiers, as used in hi-fi systems, are generally in integrated circuit form and have the advantage of reducing circuit complexity and improving the reliability of the circuits. □ see GAIN, OPERATIONAL AMPLIFIER

amplitude *n* the strength, usu measured in volts, of a signal □ see GAIN, DECIBEL, AMPLITUDE MODULATION

amplitude modulation *n* (abbr **AM**) a method of sending information by controlling the amplitude of a wave in response to changes in the frequency of the signal which represents the information. Amplitude modulation is more sensitive than frequency modulation to interference generated by lightning and other electrical discharges and it is therefore less used for high-quality radio broadcasting – compare FREQUENCY MODULATION, PULSE CODE MODULATION

analogue *adj* **1** *of information* having a continuous range of values <*sound waves are ~ signals*> **2** *of a device* responding to or displaying information as a continuous range of values <*an ~ computer*> – compare DIGITAL □ see ANALOGUE-TO-DIGITAL CONVERTER

analogue computer *n* a largely obsolete type of computer that responds to and produces continuously varying signals – compare DIGITAL COMPUTER

analogue display *n* a readout of the value of something (eg the power output of an audio amplifier) as a pointer moving over a continuous scale, or as a band of light which lengthens and shortens – compare DIGITAL DISPLAY □ see ANALOGUE MULTIMETER

analogue multimeter *n* a meter which uses analogue circuits (eg amplifiers) to measure and display, as a pointer moving over a scale, the values of the electrical quantities volts, amperes, and ohms. The analogue multimeter is becoming less popular than the digital multimeter since it costs more to make. However, it does have the advantage that changing electrical quantities can be monitored more easily than with a digital multimeter – compare DIGITAL MULTIMETER

analogue-to-digital converter *n* (abbr **ADC**) also **digitizer** a device or circuit for changing an analogue signal into a coded digital signal. ADCs enable a microcomputer or digital system to respond to information in analogue form (eg temperature, pressure, and speed). □ see DATA CAPTURE, DATA LOGGER

analytical engine *n* – see BABBAGE

AND gate *n* a decision-making building block in digital circuits which produces an output of binary 1 when both inputs are at binary 1, and an output of binary 0 when either or both inputs are at binary 0. AND gates are generally used in integrated circuit packages. □ see GATE 1

android *n* a mechanical computer-controlled robot that looks human and walks and talks <*R2-D2 was an ~ in the film Star Wars*> □ see ROBOT

angstrom unit *n* a distance equal to one ten thousand millionth of a metre (10^{-10}m) and used for measuring the wavelength of visible light *<the wavelength of red light from a light-emitting diode is about 6500 ~s>*

animated graphics *n* pictures or shapes which move across a screen under the control of a program *<adventure games make effective use of ~>* □ see SPRITE GRAPHICS

animations *n* the use of computer graphics to produce moving pictures on a VDU □ see HIGH RESOLUTION GRAPHICS

anode *n* the terminal of a device (eg a diode) towards which electrons flow – compare CATHODE □ see ELECTRODE, TERMINAL

antenna *n* a relatively complex form of aerial used in specialized radio communications systems *<a satellite tracking ~>* □ see AERIAL

antiglare surface *n* a special coating on the surface of some VDUs which makes it easier to see images by reducing reflection

anti-static packaging *n* a plastic packing material designed to get rid of static electricity and which therefore protects components (eg CMOS devices) which are especially sensitive to possible damage by static electricity □ see STATIC ELECTRICITY

APL *n* [*a* programming *l*anguage] a high-level computer language specially developed for mathematical operations. Although favoured in academic circles, APL has limited appeal for home computer users since it requires special symbols, which are not available on most keyboards, and special graphics characters.

applications program *n* a program which instructs a computer to solve a problem or to perform a specified task for a user – compare SYSTEM PROGRAM

arcade game *n* a computer game which uses a special 'display' microprocessor to display colourful, fast-moving, and exciting graphics. The display is divided into various windows for graphics and text which are held as pages in the computer's memory. The microprocessor knows which page should effectively come in front of another on the screen so that the graphics give an effect of depth and perspective. □ see VECTOR GRAPHICS, WINDOW

architecture *n* the physical arrangement and interconnections of the various parts of a microprocessor and the computer system it controls *<the ~ of a transputer>*

archive(s) *n* seldom-used information (eg legal files) on tapes and disks which act as a backup in case something happens to current information

area *n* a group of data locations in computer memory *<an operating system ~>*

argument *n* the number that a function works on (eg in Basic, SQR2 means square root of 2 where SQR is the function and 2 the argument)

arithmetic *n* any calculation involving addition, subtraction, division, or multiplication □ see ARITHMETIC AND LOGIC UNIT

arithmetic and logic unit *n* (abbr **ALU**) a section on a microprocessor chip consisting of logic gates where binary numbers are added and subtracted, and decisions are made on the results of these calculations □ see CARRY FLAG, ACCUMULATOR 1

arm *n* – see ROBOT ARM

array *n* **1** a geometric pattern of devices which form part of an imaging system <*an ~ of aerials at a radio astronomy observatory*> **2** a collection of memory locations in a computer □ see UNCOMMITTED LOGIC ARRAY, CHARGE-COUPLED DEVICE

array processor *n* also **parallel processor** a high-speed computer consisting of a number of individual microprocessors, each of which contributes the results of computations to the final result. Array processors are used for analysing the results of seismic studies in mineral exploration, for example. □ see BIT-SLICE MICROPROCESSOR, PARALLEL PROCESSING, FIFTH GENERATION COMPUTER

artificial intelligence *n* (abbr **AI**) a branch of computer science which aims to produce machines capable of exhibiting behaviour which would be considered intelligent if done by a human. Artificial intelligence was founded in the mid-1950s and its development has run parallel with the international growth in the semiconductor industry. Many industrial nations are currently to very interested in bringing together AI and microelectronics to produce a new generation of computers which are both fast and smart. □ see SOFTWARE ENGINEERING, KNOWLEDGE ENGINEERING, LISP, PROLOG, INFORMATION TECHNOLOGY, ROBOTICS, VERY LARGE SCALE INTEGRATION, FIFTH GENERATION COMPUTER

artificial satellite *n* any of a few thousand manmade objects which orbit the Earth at various heights. Artificial satellites depend greatly on microelectronics systems to carry out a variety of tasks of a military, commercial, astronomical, meteorological, and telecommunications nature. □ see COMMUNICATIONS SATELLITE, DIRECT BROADCAST SATELLITE, INTELSAT, IRAS, WEATHER SATELLITE

ASCII *n* [*A*merican *s*tandard *c*ode for *i*nformation *i*nterchange] a set of 128 characters comprising letters, numbers, punctuation marks, and symbols, each represented by a 7-bit binary word, which was introduced to facilitate the exchange of information between a computer and other data processing equipment. Most microcomputers use the ASCII code but often with slight variations to the chosen graphics symbols and some omissions from the standard set of characters. The 7-bit word associated with each character is stored in the microcomputer's ROM so that, should the computer be asked to print the character whose code is 61 (0111101 in binary), the result is the character = (ie the equals sign).

aspect ratio *n* the ratio of the width of a viewing screen (eg a VDU) to its height <*the ~ of a TV screen is about 4:3*>

assemble *vb* **1** to make up a complete computer program by bringing together subroutines **2** to translate from assembly language to machine code □ see ASSEMBLER

assembler *n* a program which converts assembly language into machine code automatically □ see INTERPRETER

assembly language *n* computer instructions which are written in mnemonic form, each mnemonic corresponding to an instruction in machine code □ see ASSEMBLER, MNEMONIC CODE

assembly listing *n* a table showing the machine code equivalent of each assembly language instruction in a computer program □ see ASSEMBLER

assign *vb* to give a value to a variable (eg in Basic, LET A = 2 means the variable A is assigned a value of 2)

astable *also* **astable multivibrator** *n* a circuit designed to produce a square or rectangular waveform. There are several ways of designing astables, which are used for generating sound effects in synthesizers, for flashing lamps and other indicators in alarm and monitoring systems, and for providing the timing signals required by microprocessors and other digital circuits. □ see CRYSTAL CLOCK

asymmetrical *also* **asymmetric** *adj, of a waveform* having unequal high and low times <*a rectangular wave has an ~ waveform*> – compare SYMMETRICAL – **asymmetrically** *adv,* **asymmetry** *n*

asynchronous *adj* of or being a communications system that sends information (eg bits, characters, or events) as it is ready. Each piece of information requires a 'start' bit and a 'stop' bit to indicate to the receiving end of the communications system when one piece has been transmitted and another starts – compare SYNCHRONOUS □ see UART

ATC – see AIR TRAFFIC CONTROL

atom *n* **1** the smallest particle of a chemical element that can exist alone or in combination with other atoms <*an ~ of silicon*> **2** a proposition in logic which cannot be broken down into other propositions

atomic number *n* the number of positively charged protons in the nucleus of an atom. The atomic number of an atom determines, to a large extent, the physical and chemical properties of an atom <*the ~ of silicon is 28*>

attack *n* – see ADSR

attenuation *n* the (usu unwanted) decrease in the strength of a signal in passing through a communications system. Attenuation has a number of causes which include the absorption of the energy of the signal by the materials through which it passes, the reflection by objects in the path of the signal, or the spreading of the signal caused by distance – **attenuate** *vb*

attenuator *n* a circuit, usu consisting of a network of resistors, which reduces the strength of a voltage, current, or resistance by a fixed ratio. Attenuators, usu in integrated circuit form, are used for calibrating measuring instruments, and in communications systems and signal generators – **attenuation** *n,* **attenuate** *vb*

attributes *n* the characteristics of a display or part of a display on a VDU. Attributes are usu stored in a single byte of RAM for each screen location. Each bit or group of bits within that byte stores one of these attributes which can be specified using the commands 'bright', 'flash', etc.

audio *adj* of or being oscillations (eg sound waves, mechanical vibrations, or electrical oscillations) having frequencies in the range of human hearing <*an ~ system*> □ see ACOUSTIC

audio *n* the transmission, reception, and reproduction of sound by electronic means

audio amplifier *n* an electronic device or system for amplifying electrical signals in the audio range (ie from 20 hertz to 20 kilohertz). An audio amplifier in a hi-fi system usu comprises a preamplifier, tone controls, and power amplifier. □ see HARMONIC DISTORTION, PUSH-PULL AMPLIFIER

audioconferencing *n* a meeting between people in different places who communicate by telephone and which usu involves the sending of printed material through facsimile □ see TELECONFERENCING

audio frequency *adj* (abbr **AF**) *of waves* having a frequency in the range of human hearing. Audio frequency is usu taken to refer to the range 20 hertz to 20 kilohertz <*an ~ amplifier*>.

audio oscillator *n* a device for producing signals covering the range of human hearing (approximately 20 hertz to 20 kilohertz). Audio oscillators are frequently used for testing the performance of amplifiers. □ see WAVEFORM GENERATOR

automatic *adj* acting by itself <*an ~ telephone exchange*> – **automatically** *adv* □ see AUTOMATION

automatic answering service *n* a telephone which, on receiving a call, automatically provides the caller with a prerecorded telephone message

automatic frequency control *n* (abbr **AFC**) a circuit used in frequency modulation radio receivers to ensure that a station keeps in tune

automatic gain control *n* (abbr **AGC**) a function, usu included in the design of amplifiers, which ensures that both weak and strong signals (eg from a microphone) are amplified to a fixed level

automation *n* the operation of any manmade device or system which takes the place of human operators (eg the welding and assembly of cars on a production line by robots) □ see ROBOTICS, CYBERNETICS, ARTIFICIAL INTELLIGENCE

auto-repeat *adj* of or being a key on the keyboard of a computer or wordprocessor that provides automatic repetition of a symbol when held down

avalanche breakdown *n* the sudden rise in current through a reverse-biased pn junction caused by electrons having sufficient

energy to dislodge further electrons from their parent atoms. Some
Zener diodes used for voltage regulation are designed to make use of
avalanche breakdown, but rectifier diodes used in mains-operated
DC power supplies are usu destroyed by it. □ see ZENER DIODE

avionics *n* electronic devices and systems (eg automatic pilot) used
in aviation □ see RADAR

AWAC [*a*irborne *w*arning *a*nd *c*ontrol] a powerful radar system
carried aboard a Boeing aircraft to maintain a continuous watch on
possibly hostile missiles and aircraft. Aircraft fitted with AWAC
have a mushroom-shaped dome on their fuselage. □ see RADAR

B

B 1 the symbol for byte □ see BYTE **2** the symbol used in the hexadecimal system for the decimal number 11 (eg the hexadecimal number 7B equals the decimal number $7 \times 16 + 11 = 123$) □ see HEXADECIMAL

Babbage, Charles (1792-1871). Designed an 'analytical engine', the world's first (mechanical) computer. It could be 'programmed' from punched cards, digits could be stored and added and subtracted, and results could be printed out. However, it didn't work because its component parts could not be machined accurately enough. □ see ADA

back emf *n* a usu high voltage (ie an electromotive force) produced by a changing magnetic field round an inductor (eg a relay coil) which tends to oppose the change in current responsible for the magnetic field. Precautions are usu taken to ensure that the back emf does not damage other circuit components, esp semiconductor devices, and one method of protection is to use a diode to bypass the back emf. □ see INDUCTOR, LENZ'S LAW

back end *n* **1** that part of a computer system providing additional facilities <*a central heating control board is the ~ of a computer*> **2** that part of a communications system sending processed signals somewhere else <*the ~ of a radio is its speaker*> – compare FRONT END – **back end** *adj*

background colour *n* the colour chosen by a programmer for the screen of a VDU on which information is written in another colour – compare FOREGROUND COLOUR

background noise *n* any continuous interference on a communications channel <*hiss is ~ on a radio receiver*> □ see NOISE, WHITE NOISE

background processing *n* the automatic running of lower priority computer programs when higher priority programs are not using the resources of the computer – compare FOREGROUND PROCESSING

backing store *n* any program store or data store (eg floppy disk) which is external to a computer and whose contents do not become erased, like programs stored in RAM, when the computer is switched off – compare MAIN MEMORY

backplane *n* a connector system plugged into a computer to extend its capabilities. For a microcomputer, a backplane usu consists of a printed circuit board carrying edge connectors into which other circuit boards can be plugged. □ see MOTHERBOARD, PERIPHERAL

backspace *vb* **1** to move the carriage on a printer one space to the right **2** to move a cursor on a VDU one space to the left **3** to reverse a magnetic tape to the beginning of the previous block of data – **backspace** *n*

backup *adj* duplicating the function of something in case of emergencies <*a ~ computer on a Shuttle flight*> – **backup** *n*, **back up** *vb*

badge reader *n* a security device (usu connected to a computer terminal) that reads coded information (eg a magnetic stripe) on a badge to identify the person to whom the badge belongs □ see SMART CARD

balanced power supply *n* a power supply which provides nominally equal positive and negative voltages above and below 0 volts (ie ground voltage). A balanced power supply is useful when working with some types of integrated circuit, esp op amps and analogue-to-digital converters <*a ±15V ~*>.

band *n* a range of frequencies <*the FM ~ on a radio receiver*> □ see BANDWIDTH

bandolier *n* a continuous double strip of paper tape between which resistors, capacitors, diodes, etc are fixed by their wires for easy packaging and shipment – **bandoliered** *adj*

band pass filter *n* a circuit which allows a narrow band of radio frequencies to pass through it. Band pass filters are used to select stations in very high frequency radio receivers. □ see LOW-PASS FILTER, HIGH-PASS FILTER

bandwidth *n* the frequency range of the messages handled by an information processing system. Bandwidth is important when considering the 'volume' of information transmitted along a communications channel. As a general rule in a digital system each bit of data to be transmitted per second requires a bandwidth of 2 hertz. Thus a bandwidth of 600 hertz is needed to transmit 300 bits per second; normal speech along a telephone line needs a minimum of 3000 hertz; and a colour TV picture needs a bandwidth of about 8 million hertz to be transmitted. The capacity of a transmission system is usu expressed in terms of the number of telephone conversations it can carry. Thus the largest capacity coaxial cable has a bandwidth of 500 megahertz and can therefore carry about 167 thousand telephone conversations. And a hair-thin glass fibre can carry about 10 thousand telephone conversations. □ see TELEPHONE, FREQUENCY, OPTICAL COMMUNICATIONS, CABLE TV

bank *n* **1** – see DATABANK **2** a group of items connected side-by-side <*a ~ of switches*>

bar coding *n* a method of labelling items in shops, offices, factories, laboratories, and warehouses with a sequence of thick and thin lines representing information which can be read by passing a penlike bar code scanner over the label to enter data into a computer system. The scanner may be passed over a label in either direction depending on the orientation of the product. In a typical set of lines, the extreme right and left lines 'tell' the scanner the direction in which the lines are being read. The coding of the bars is interpreted by a read-only memory in a computer or hand-held terminal linked to the scanner. In a shop, the computer not only stores the latest price of the goods displayed to the customer, but also produces the bill and updates a database enabling goods to be reordered when they fall below a critical stock level. □ see POINT-OF-SALE TERMINAL

Bardeen, Dr John. One of the three-man team which won the Nobel Prize in 1956 for the invention of the transistor in 1947 at the Bell Telephone Laboratory, USA. The other two men were Dr William Shockley and Dr Walter Brattain. □ see TRANSISTOR

bargraph display *n* an instrument readout consisting of a row of light-emitting diodes, a number of which are lit to indicate the value of some measurement (eg a car's speed or the audio power output of a hi-fi amplifier) □ see LIGHT-EMITTING DIODE, SEVEN-SEGMENT DISPLAY

baroque technology *n* the extravagant and sometimes needless use of sophisticated electronic equipment to enhance the performance of something. Baroque technology is exhibited in some makes of hi-fi equipment and cars, esp in the overuse of digital displays and voice synthesis, and it is also seen in weapons systems. Manufacturers of tanks, fighters, and rocket launchers are exploiting promising state-of-the-art electronic devices and systems in order to gain orders for their equipment – what some people call 'the all-singing, all-dancing weapons system'.

bar printer *n* a printer (eg a mechanical typewriter) which has characters embossed on separately movable bars

base *n* **1** *also* **radix** the number of digits in a number system <*the binary number system has a ~ of 2*> **2** one of the three terminals of a bipolar transistor into or out of which the controlling current flows

base address *n* the lowest numbered location in a computer storage device □ see ADDRESS

Basic *n* [*b*eginners *a*ll-purpose *s*ymbolic *i*nstruction *c*ode] a high-level computer language specially constructed to make computers easy to use by nonspecialists. Basic is the language most computer users use to 'talk' to their computers. □ see HIGH-LEVEL LANGUAGE

batch *n* a group of records which a computer processes as one lot – **batch** *vb*

batch processing *n* a method of processing data in which transactions (eg updating investors' deposit accounts) are collected together in batches and run on a computer when time permits

battery *n* a set of cells (eg carbon-zinc dry cells) usu connected in series which produces a direct current <*a radio ~*> □ see CELL

battery backup *n* the provision of electric power from a battery mounted on a circuit board or built into equipment in the event of failure of the main supply. Battery backup is useful for ensuring that data is not lost from a volatile memory in the event of power failure to a microcomputer. It is also used in burglar alarms and other vital mains-operated equipment.

baud *n* a unit for measuring the speed at which groups of binary digits are transmitted one after the other (ie serially) in digital communications systems (eg between one computer and another along a telephone line). If the binary data is transmitted one bit at a time, 100 baud equals 100 bits per second. However, the data might comprise two-bit groups in which case 100 baud equals a bit rate of 200. Typical transmission rates in baud are 110, 300, 600, 1200, 2400 and 4800

BCD – see BINARY-CODED DECIMAL

BCD counter *n* [binary-coded decimal counter] a binary counter which, on receiving pulses (counts), produces the first ten binary codes, (0000) to (0101), one after the other before resetting to (0000) and beginning the count again □ see BINARY COUNTER

beam leads *n* flat metal bars attached to some semiconductor devices (eg an integrated circuit audio amplifier) to act as both terminals and heat sinks – **beam lead** *adj* □ see HEAT SINK

beat frequency *also* **beat note** *n* the frequency which results from adding together two waves of slightly different frequencies. Beat frequency is equal to the difference between the two frequencies. Thus two sound waves of frequencies 220 hertz and 225 hertz produce a beat frequency of 5 hertz.

beat note *n* BEAT FREQUENCY

bel *n* ten decibels □ see DECIBEL

benchmark *n* a standard test given to different computers to compare their speed of operation and data processing capabilities

bias *n* the provision of a current or voltage in a circuit to make the circuit function properly <*transistor* ~> – **bias** *vb*, **bias** *adj*

bicolour LED *n* a light-emitting diode which produces one colour (eg red) when current flows through it one way, and another colour (eg green) when current flows through it in the opposite direction. A bicolour LED is useful in some circuits for indicating a 'stop' or 'go' situation. □ see LIGHT-EMITTING DIODE

bidirectional *adj* **1** *of a communications channel* allowing information to pass through it in either direction <*a* ~ *data bus in a computer*> **2** *of a printer* printing the first line from left to right, the next line from right to left, and so on in order to save on printing time – compare UNIDIRECTIONAL

BIFET *n* [bipolar-compatible field-effect transistor] a junction-gate field effect transistor which is combined with bipolar transistors in some types of integrated circuits (eg operational amplifiers). The combination reduces input current and makes the device suitable for designing electronic instruments.

BIMOS *n* a metal-oxide field-effect transistor which is combined with bipolar transistors in some types of integrated circuits (eg operational amplifiers). The combination reduces input current to a very low level and makes the device particularly suitable for designing electronic instruments.

binary *adj* of or belonging to a system of numbers with a base of 2 <*the binary number system uses the* ~ *digits 0 and 1*>

binary arithmetic *n* arithmetic performed with binary numbers. The addition of binary numbers is the basis of all binary arithmetic, esp that performed by the arithmetic and logic unit of a microprocessor. □ see ONE'S COMPLEMENT ARITHMETIC

binary cell *n* the basic unit of computer memory, capable of storing binary 1 or 0

binary code *n* numbers expressed as groups of binary digits as used by computers □ see BINARY NUMBER, BINARY-CODED DECIMAL

binary-coded decimal *n* (abbr **BCD**) any of the 10 binary numbers equivalent to the decimal numbers 0 to 9 <*0110 is the ~ number equal to the decimal number 10*>. Digital circuits for driving seven-segment displays (eg in a digital watch) use binary-coded decimal numbers.

binary counter *n* any of a variety of digital integrated circuits used for counting events (eg the pulses produced by a quartz crystal in a digital watch) represented by the binary digits 1 and 0 □ see RIPPLE COUNTER

binary digit *n* either of the digits 0 and 1 used to express a number (eg 0110) in the binary number system □ see BIT

binary divider *n* any device (eg a prescaler), usu in integrated circuit form, used for dividing a frequency by some factor. For example, in digital clocks and watches a divide-by-60 binary divider is used to convert one-second pulses into one-minute pulses. Binary dividers, like binary counters, are made from flip-flops. □ see BINARY COUNTER, FLIP-FLOP

binary number *n* a number (eg 1011) in the number system which has a base of 2. The right-most digit has a weighting of 2^0 (=1); the next digit a weighting of 2^1 and so on. Thus the number above has a decimal equivalent to 11. Binary numbers can be added, subtracted, and multiplied according to the normal rules of arithmetic. Electrical signals in the form of binary numbers are used in all digital circuits (eg calculators and computers).

binary storage *n* binary numbers held in a memory device as combinations of 0s and 1s. Various physical principles are used to store the data depending on the construction of the device. For example, a static random-access memory stores binary numbers using the on-off states of transistor flip-flops. □ see RANDOM-ACCESS MEMORY, MAGNETIC BUBBLE MEMORY, FLIP- FLOP

biochip *n* a silicon chip implanted in the body to make an on-the-spot check of chemical activity. A biochip is being designed for operating alongside a pacemaker so that the pacemaker can respond to the presence of adrenalin in the blood just as a healthy heart does. The long-term aim is to communicate directly with the brain by implanting a biochip in it – a prospect dubbed bolt-on intelligence. A silicon chip is at the heart of a biochip but it is protected by a layer of special glass or plastic through which only those chemicals it is designed to detect can pass. Between the glass and the chip is a layer which converts the chemicals into electrical impulses which are interpreted by the chip itself □ see BLIND AIDS, SILICON CHIP

bioelectronics *n* the implantation of electronic devices in the body to help people lead fuller lives. The term also embraces the development of electronic devices which use living tissue (eg memory devices based on living plant cells). Strictly, bioelectronics includes only those applications where there is intimate contact between the body tissue and manmade electronic devices (eg an artificial electronic eye fitted into the eye socket of a blind person and coupled directly to the optic nerve). Thus the conventional implanted heart pacemaker is not an application of bioelectronics. □ see BIOCHIP, PACEMAKER, BLIND AIDS, DEAF AIDS

bionic *adj* [*bio*logical electro*nic*] of or being an electronic device which has the characteristics of living organisms □ see BIOELECTRONICS, ANDROID

biotelemetry *n* the remote measurement of the life functions (eg heart beat and respiration) of animals using radio and often involving the use of communications satellites as relay stations. Biotelemetry is used to monitor the activity of many animals, from migrating moose to feeding fish <~ *of astronauts aboard the Shuttle*> □ see TELEMETRY

bipolar *adj of a semiconductor device (eg a transistor)* made from p-type and n-type semiconductors – compare UNIPOLAR □ see BIPOLAR TRANSISTOR

bipolar transistor *n* a three-terminal device made from p-type and n-type semiconductors which is used for switching and amplifying electrical signals in circuits. A bipolar transistor can be obtained as a single (discrete) component or used along with many others integrated on a silicon chip. There are two types of bipolar transistor, the npn type and the pnp type. The size of the current flowing between its emitter and collector terminals is controlled by the current flowing into or out of its base terminal. In integrated circuits, the bipolar transistor can switch current on and off much faster than a field-effect transistor.

bistable *adj* having two stable states <*a ~ multivibrator*> □ see FLIP-FLOP

bistable *also* **bistable multivibrator** *n* FLIP-FLOP

bit *n* a binary digit (eg 0 or 1), the basic unit of information in computer and other digital systems. Generally thought to be a contraction of *b*inary dig*it*, it is quite possible that a bit is just simply an extension of its common meaning 'a small piece of something'. □ see BYTE, NIBBLE, WORD

bit density *n* the number of binary digits (ie 1s and 0s) stored per unit length, area, or volume of a storage medium

bit pad *n* – see DIGITIZER 1

bit pattern *n* the sequence of 0s and 1s in a binary word <*the ~ of the nibble 1001*>

bit rate *n* the speed, in bits per second, at which binary data is transferred along a communication channel <*the ~ from a space probe's antenna*> □ see BAUD

bit-slice microprocessor *n* a microprocessor made up of several connected integrated circuits all based on fast-acting bipolar transistors. Integrating complex bipolar transistor circuits on a silicon chip is difficult because of their higher heat output compared with metal-oxide semiconductor transistors. But by bringing together a number of smaller bipolar integrated circuits, fast processing speeds can be obtained with the overall complexity required. Almost all new computers in recent years have been based on the bit-slice microprocessor, and the same principle is being proposed for fifth generation computers.

black box *n* any electronic component or equipment which is looked at from the point of view of what it does rather than how it works. Thus the user of a hi-fi system would regard the amplifier as a black box when he/she connects it to a turntable, power supply, and speakers and operates its controls to listen to music. □ see FLIGHT RECORDER

blank *adj* not containing information <*a ~ tape*>

blind aids *n* microelectronic devices for helping blind people live a fuller life. The talking calculator uses integrated circuits which synthesize speech. Hand-held 'torches' which provide audible information about the nearness and shapes of objects ahead are in use. The long-term aim is to develop a video camera small enough to be implanted in a blind person's eye socket to generate electrical signals for stimulating the brain's 'seeing area' (the visual cortex).

blink *vb* to draw attention to something (eg a cursor on a VDU) by switching it on and off

block *n* the standard unit for allocating and transferring data on magnetic disk or tape

block diagram *n* a diagram which shows how the building blocks of an electronics system are connected together <*a ~ of a radio receiver*> □ see BUILDING BLOCK, SYSTEM, BLACK BOX

blocking *adj* stopping the transmission of signals <*a ~ diode*> – **block** *vb* □ see COUPLING CAPACITOR

blocking capacitor *n* COUPLING CAPACITOR

BNC *n* [bayonet connector] a connector for joining coaxial cables together

board *n* – see PRINTED CIRCUIT BOARD, MOTHERBOARD

Bode plot *n* a graph showing how the voltage gain and phase shift of an amplifier, esp an operational amplifier, vary with the frequency of the signal passing through it. A Bode plot provides the circuit designer with information about the stability of the amplifier at high frequencies. □ see BANDWIDTH

Boole, George (1815-64). Founder of Boolean algebra, the mathematical logic used for predicting the behaviour of digital circuits (eg those in computers). □ see BOOLEAN ALGEBRA

Boolean algebra *n* a shorthand way of writing complex combinations of logical statements which are either true or false. Boolean algebra was developed by George Boole, a 19th-century mathematician, and it has subsequently become a useful way of analysing and predicting the behaviour of digital circuits which deal with on and off signals, since true can be regarded as logic 1 (eg an on signal) and false as a logic 0 (eg an off signal). □ see GATE 1, TRUTH TABLE, BOOLE

Boolean operator *n* a logic function (eg AND), or a combination of logic functions □ see BOOLEAN ALGEBRA

boot *n* the action of loading an operating system from a magnetic disk into a computer's memory – **boot** *vb* □ see BOOTSTRAP, COLD BOOT, WARM BOOT

booting DOS *n* – see DISK OPERATING SYSTEM

¹**bootstrap** *vb* to increase the input resistance of an amplifier by making a resistor appear to have a higher resistance than it really has. This circuit design technique is used to make sure that signals from a high resistance source (eg a crystal microphone) are amplified more efficiently.

²**bootstrap** *also* **bootstrap loader** *n* **1** a short machine code routine, usu stored in a computer's ROM, which is activated when the computer is first switched on and which finds the operating system on disk or tape, transfers it byte-by-byte into a fixed area of RAM, and then hands over control to the new program. The bootstrap gets its name from the fact it pulls the main program into the computer's memory as if by its 'own bootstraps'. **2** BOOT – **bootstrap** *vb* □ see DISK OPERATING SYSTEM

boron *n* a trivalent metallic element of atomic number 5 which is allowed to diffuse, in the form of a gas, into microscopically small areas of a silicon chip during the making of an integrated circuit □ see P-TYPE, VALENCY, PHOTOLITHOGRAPHY

boule *n* also **ingot** a very pure single crystal of silicon in the form of a solid cylinder typically 50 to 150 millimetres long. A boule is obtained by slowly pulling the growing crystal from a bath of pure molten silicon and is the starting point in the complex chain of operations for making integrated circuits on silicon chips. □ see PHOTOLITHOGRAPHY, WAFER

BPS [*bits per second*] – see BAUD, BIT RATE

braid *n* – see DESOLDER BRAID, COAXIAL CABLE

branch *n* an instruction (eg Gosub) which tells a computer to go to another line in a program □ see JUMP

Brattain, Walter – see BARDEEN

breadboard *n* a system for the speedy assembly and testing of prototype circuits which does not involve soldering. A breadboard usu consists of a baseboard into which wires are plugged. Interconnections between components are made both automatically, via connections built into the board, and manually by jump leads.

break *vb* **1** to stop transmitting or receiving a message **2** to halt a program by pressing a break key – **break** *n*

breakdown diode *n* ZENER DIODE

breakdown voltage *n* the sudden and usu destructive current which flows through a semiconductor diode when the reverse voltage across it exceeds a specific value – compare WORKING VOLTAGE □ see ZENER DIODE, AVALANCHE BREAKDOWN

break key *n* a key on a computer keyboard which stops a program running and returns control to the programmer. The stop key and escape key have the same function.

breakpoint *n* a place in a computer program where execution has to be halted □ see INTERRUPT

bridge rectifier *n* a single 4-pin package containing 4 rectifier diodes, used in power supplies for the full-wave rectification of alternating to direct current □ see HALF-WAVE RECTIFIER, FULL-WAVE RECTIFIER

broadband *n* WIDEBAND

¹broadcast *adj* of or being radio or television signals transmitted in all directions <~ *receiver for radio programmes*> □ see DIRECT BROADCAST SATELLITE

²broadcast *vb* to transmit the same message to a widely scattered audience <*to ~ the Queen's Christmas message*> – **broadcast** *n*

brownout *n* a reduction (ie not a blackout) in the mains power supply voltage which might jeopardize the operation of something. A microcomputer can withstand a brownout if it has battery backup to retain data in its memory.

browsing *n* (idly) looking through computer files in the hope of finding something interesting; 'software loitering' – **browse** *vb*, **browser** *n*

bubble memory *n* MAGNETIC BUBBLE MEMORY

bucket brigade *n* a memory device (eg magnetic bubble memory) in which data is continuously circulated □ see CHARGE-COUPLED DEVICE, MAGNETIC BUBBLE MEMORY

buffer *n* **1** a device connected between two other devices in order to strengthen a signal. A microprocessor uses buffers to enable it to send signals to other parts of the computer system. **2** a device in an information processing system which receives data from a device operating at one speed and releases it to another device operating at a slower speed. A microcomputer uses a buffer to release characters from its memory to a printer at a speed the printer can handle. □ see BUS DRIVER, FIFO **3** an area of a computer's memory specially set aside for storing blocks of data temporarily. A buffer is used, for example, in a wordprocessor program when blocks of text are moved from one place to another. – **buffer** *vb*

bug *n* **1** anything which prevents a program, device, or circuit from working properly <*a software ~*>. The effect of a bug is often tolerated in which case it is known as a feature. □ see HOPPER, DEBUG **2** a small and inconspicuous electronic device for listening undetected to private conversations. Most bugs use their own power supply but others draw on power from the device (eg a telephone handset) to which they are attached or are energized by the sound waves produced when people talk. The design of bugs has benefitted by progress in the field of microelectronics. – **bug** *vb*

buggy *n* **1** a remote-controlled wheeled or tracked vehicle for exploring the surface of a planet. This type of buggy is controlled by an on-board computer which can be overridden by commands direct from Earth. Communication with the buggy is either direct from Earth or by relaying data via an artificial satellite orbiting the planet. □ see INTERPLANETARY PROBE, VIKING **2** a small computer-controlled motorized vehicle which provides an entertaining way to learn programming skills related to controlling devices. The buggy carries on-board electronics for operating its motors (eg stepping motors), for sensing obstacles, and, when the buggy is used for drawing diagrams, for raising and lowering a pen. The computer provides digital signals for activating the on-board electronics, and for receiving signals from the sensors. The signals flow to and from the buggy by cable, or by infrared or radio. □ see TURTLE, ROBOT

building block *n* a circuit (eg a flip-flop) that has a particular function and can be used with others to produce a more complex function (eg a computer memory) □ see BLACK BOX, SYSTEM

bulk storage *n* MASS STORAGE

bus *n* an electrical route along which data flows between two or more devices in an electrical system. A microcomputer has several busses (eg a data bus) connecting the microprocessor to other devices.

bus driver *n* a specially designed integrated circuit in a computer system which ensures that data from the microprocessor is in a form suitable for transmission along a data bus □ see DATA BUS, BUFFER

business computer *n* a personal computer or a microcomputer which is primarily intended for business applications. Many modern business computers are portable and designed to fit inside an attaché case; they often have a built-in LCD or electroluminescent screen; they are very user-friendly, employing a mouse or touchscreen facility so that the average person finds it easy to use; and business software (eg a spreadsheet) is usu supplied by the company marketing the computer.

bus network *n* a popular communications system between computers in which data is passed directly from one computer to another <*Econet uses a* ~> – compare STAR NETWORK, RING NETWORK

buzzer *n* an audible warning device usu consisting of an oscillator which drives a small loudspeaker on and off at low frequency

buzzword *n* a jargon word for any of the many jargon words associated with microelectronics. Buzzword was coined by Honeywell's publicity department in 1960 when they developed a simple game called 'Buzzword Generator'. This game centred on three columns of ten words each. The first column contained adjectives and the second two columns, nouns. All the player had to do was to think of a three digit number (eg 285) and then look up the corresponding words in the columns. In this way, a (usu) meaningless phrase such as 'digitized graphic facility' could be found with which to baffle and 'impress' colleagues and friends.

bypass capacitor *n* a capacitor connected in a circuit to divert an unwanted signal (eg a carrier wave in a radio receiver) to the circuit ground or earth □ see CAPACITOR

byte *n* (symbol **B**) a group of binary digits. Originally equal to the number of bits required to encode a single character, the byte is now generally used for an 8-bit word which is moved around as a single unit of data in computer systems. □ see NIBBLE, GULP, HEXADECIMAL

C

C 1 the symbol for electrical capacitance □ see CAPACITANCE **2** the symbol used in the hexadecimal system for the decimal number 12 (eg the hexadecimal number C4 equals the decimal number $12 \times 16 + 4 = 196$) □ see HEXADECIMAL

cable *n* a group of electrical conductors or optical fibres running side-by-side in a protective sheath □ see CABLE TV, OPTICAL COMMUNICATIONS, COAXIAL CABLE

cable TV *n* (abbr **CATV**) the distribution of TV programmes to viewers by means of underground cables. Though well-established in the USA, cable TV is still under review in the UK. As well as offering a very wide choice of programmes, cable TV can be used for shopping and banking from home, accessing library resources, voting on local affairs, entering competitions, and other interactive purposes. □ see COAXIAL CABLE, VIEWDATA, DIRECT BROADCAST SATELLITE, TELESHOPPING, TELEBANKING

CAD – see COMPUTER-AIDED DESIGN, CAD/CAM

CAD/CAM [*computer-aided design/computer-aided manufacturing*] computer-aided design and computer-aided manufacturing considered together as a single concept; the (increasing) use of computers both in the design and the manufacture of products (eg cars) □ see COMPUTER-AIDED DESIGN, COMPUTER-AIDED MANUFACTURING

cadmium sulphide photocell *n* a light-sensitive device used for measurement (eg in camera lightmeters) and control (eg in automatic street lights) whose resistance falls with increasing light intensity □ see LIGHT-DEPENDENT RESISTOR

CAI – see COMPUTER-AIDED INSTRUCTION

CAI/CAL [*computer-aided instruction/computer- aided learning*] – see COMPUTER-AIDED INSTRUCTION, COMPUTER-AIDED LEARNING

CAL – see COMPUTER AIDED LEARNING

calculator *n* – see ELECTRONIC CALCULATOR

Call *n* a keyword within a program which instructs a computer to start on another routine (eg go to a machine code routine from a Basic program). Different computers use variations of this command (eg Gosub, Proc, and Usr)

CAM – see COMPUTER-AIDED MANUFACTURING, CAD/CAM

camera *n* – see ELECTRONIC CAMERA, VIDEO CAMERA

capacitance *n* (symbol **C**) the ability of a component, esp a capacitor, to store electrical charge – **capacitive** *adj,* **capacitively** *adv* □ see CAPACITOR, FARAD

capacitive reactance *n* – see REACTANCE

capacitor *n* an electrical component designed to store electricity. Capacitors are widely used in circuits for producing time delays and electrical oscillations. A capacitor is made of an insulating material (eg polyester) sandwiched between two metal plates that form its terminals. □ see FARAD, TIMER, OSCILLATOR, TUNED CIRCUIT

capacity *n* **1** the maximum amount of memory space in a storage medium, measured in kilobytes or megabytes □ see KILOBYTE, MEGABYTE, MEMORY **2** the value of the capacitance of a capacitor □ see CAPACITOR, FARAD

capstan *n* a wheel in a cassette recorder or similar tape drive device which moves the tape

card *n* **1** PUNCHED CARD **2** a printed circuit board which is plugged into a main circuit board to increase the capacity and function of a computer system <*an interface* ∼>

carriage *n* the part of a printer that holds the platen and moves from right to left as printing takes place □ see PLATEN

carrier *n* CARRIER WAVE

carrier wave *n* (abbr **CW**) also **carrier** a relatively high frequency wave (eg a radio wave) by which a message (eg music) at a lower frequency is sent along a communications channel □ see MODULATE

carry *n* CARRY FLAG

carry bit *n* CARRY FLAG

carry flag *also* **carry, carry bit** *n* a flag (ie a 1 or 0) used by the arithmetic and logic unit in a microprocessor when it is doing addition of binary numbers. If the result of doing an addition of two numbers is too large to fit into the accumulator, the carry flag is set to a 1 to show that there is an overflow. □ see FLAG, ARITHMETIC AND LOGIC UNIT, MICROPROCESSOR

cartridge *n* a specially packaged self-contained memory chip or magnetic tape used for storing programs (eg an adventure game) and capable of being plugged or slotted into a special socket on some microcomputers □ see ROM CARTRIDGE, STRINGY FLOPPY

cartridge tape *n* STRINGY FLOPPY

cashless society *n* the predicted widespread use of 'electronic' credit cards to pay for goods and services □ see TELESHOPPING, TELEBANKING, SMART CARD

cassette *n* a package containing spools round which magnetic tape moves for recording and playing back computer programs or other information □ see FLOPPY DISK, HARD DISK

CAT – see COMPUTERIZED AXIAL TOMOGRAPHY

catalogue *n* any list of information (eg program titles) stored on tape or disk; *esp* DIRECTORY – **catalogue** *vb*

cathode *n* the terminal of a device (eg a diode) away from which electrons flow – compare ANODE □ see ELECTRODE, TERMINAL

cathode-ray oscilloscope *n* (abbr **CRO**) also **oscilloscope** a test and measurement instrument for showing the patterns of electrical waveforms and for measuring their frequency and other characteristics. The cathode-ray oscilloscope is an indispensable aid to circuit designers and test engineers, esp those involved in the design and maintainance of computer and telecommunications systems. Although most oscilloscopes are based on the cathode-ray tube, recent designs make use of a large-screen liquid crystal display for increased portability and lower power consumption. □ see CATHODE-RAY TUBE

cathode rays *n* a stream of fast-moving electrons carrying a negative charge and used in TV tubes, oscilloscopes, and radar displays to produce images on a screen □ see CATHODE-RAY TUBE, VDU, ELECTRON, PHOSPHOR

cathode-ray tube *n* (abbr **CRT**) a glass tube (eg a TV tube) in which a beam of electrons moves in a vacuum to 'draw' information on a screen. The electron beam is produced by an electron gun and it is focussed and deflected by coils and/or electrodes. □ see TELEVISION RECEIVER, VDU, THERMIONIC EMISSION, PHOSPHOR

CATV – see CABLE TV

CB radio *n* [citizen *b*and radio] a two-way radio communication system used by the general public. CB radio was legalized in the United Kingdom in 1982 on two FM bands, 27 megahertz (MHz) and 934MHz. Forty channels separated by 10 kilohertz are available on the 27MHz band, and 20 channels separated by 5 kilohertz on the 934MHz

CCD – see CHARGE-COUPLED DEVICE

CCTV [*c*losed-*c*ircuit *TV*] – see CLOSED-CIRCUIT

CDA – see CURRENT DIFFERENCING AMPLIFIER

CD ROM *n* [compact *d*isk *ROM*] a disk on the surface of which are microscopically small pits for storing computer data which can be read using laser light. This form of data storage is identical to that used for compact audio disks, but the technique is not yet fully accepted for storing computer data. The CD ROM offers 500 to 1000 times as much storage capacity as floppy disks, equal to over 500 megabytes of data or 120 thousand A4 pages of text. □ see COMPACT DISK, FLOPPY DISK, OPTICAL MEMORY

Ceefax *trademark* – see TELETEXT

cell *n* **1** a subdivision of a computer memory which stores one unit (usu one bit) of data □ see MEMORY **2** a device that produces electricity by chemical action and is the building block from which batteries are made **3** – see CELLULAR RADIO

cellular radio *n* a reliable and fast computer-controlled communications system for users of in-car telephones. Areas of a city, town, or region are divided into radio transmission zones known as cells. Each cell has its own transmitter/receiver, usu on a high building. A cluster of cells is controlled by a central computer so that calls from a car are fed to the public telephone system or to another group of cells. A scanning system keeps track of the signal level from a vehicle and the computer automatically transfers to a transmitter/receiver in another cell so that the driver does not notice the passage from one cell to another. Cellular radio can easily handle 50 thousand calls per hour compared with 15 hundred calls using conventional shortwave radio, and it is generally regarded as the most important innovation in telephony since the invention of the telephone in the 1870s.

central processor (unit) *also* **central processing unit** *n* (abbr **CPU**) the principal operating and controlling part of a computer. In a microcomputer, the CPU is a microprocessor. □ see MICROPROCESSOR

Centronics interface *trademark* – used for a circuit in some microcomputers which transfers data to and from a peripheral (eg a printer) one byte at a time – compare RS-232 INTERFACE □ see PARALLEL INPUT-OUTPUT

ceramic filter *n* a device in VHF radios and other communications equipment which allows a narrow band of radio frequencies to pass through it. A ceramic filter consists of a thin plate of ceramic material (eg lead zirconium titanate) which has a thin film of metal on each face so that it vibrates at a characteristic frequency when an alternating voltage is applied to it. □ see CRYSTAL CLOCK, PIEZOELECTRICITY, BAND PASS FILTER

chain *n* a sequence of events linked together so that the completion of one event begins the next <*a ~ of programs on a disk*> – **chain** *vb*

channel *n* **1** that part of a communications system along which information travels from a source to a destination <*a telephone line is a ~*> **2** that part of a computer system which performs input/output functions **3** the conducting path between the source and drain terminals of a field-effect transistor □ see PATH, LINK

channel width *n* the number of bits handled simultaneously by a communications channel <*a ~ of 8 bits*> □ see BANDWIDTH

character *n* any letter, number, or symbol which can be produced on a keyboard for display on a VDU □ see ASCII, CHARACTER GENERATOR

character cell *n* – see CHARACTER GENERATOR

character generator *n* a special chip in a computer which produces a character on a VDU when a key is pressed. A character generator consists of a matrix of character cells each holding a graphic representation of a character. □ see ASCII

characteristic *n* a graph showing the particular way a device behaves electrically <*a diode ~*>

character set *n* a group of characters (eg the alphabet) suitable for transmission from one place to another □ see ASCII

characters per second *n* (abbr **CPS**) the speed at which (slow) devices (eg teletypewriters) handle or transmit characters □ see BAUD

¹charge *vb* to store electricity in a capacitor, cell, or battery
²charge *n* **1** the amount of electricity stored in a capacitor, cell, or battery **2** a basic property of matter that occurs in separate natural units and is considered as negative (eg when belonging to an

electron) or positive (eg when belonging to a proton) □ see
COULOMB, ELECTRON, PROTON

charge-coupled device *n* (abbr **CCD**) a data storage device
consisting of an array of metal-oxide semiconductor cells on a silicon
chip. Each cell stores a small electric charge representing a bit of
information which can be moved through the memory by means of
electrodes formed on the surface of the chip. CCDs are being
increasingly used in television cameras for storing images, esp in the
field of robotics, since a CCD can recognize not just black and white
but also a range of grey tones. The CCD is about 30 times more
sensitive than a photographic plate so it is used in astronomy for
taking pictures of faint celestial objects in minutes rather than in
hours. □ see MAGNETIC BUBBLE MEMORY

chart recorder *n* an electronic device showing how some quantity
(eg atmospheric pressure) varies with time □ see DATA LOGGER

chassis *n* the metal frame on which circuit boards and components
are mounted <*the ~ of a television set*> □ see ¹EARTH

check bit *n* any binary digit (ie 0 or 1) that shows the result of a
processing operation in a computer or communications system □ see
PARITY, FLAG

chip *n* – see SILICON CHIP

chip count *n* **1** the number of integrated circuits required to perform
a particular function <*the ~ of a television*> **2** the number of
integrated circuits on a particular circuit board

chip head *n* someone crazy about microelectronics and computing; a
'computer nut' □ see COMPUTERNIK, HACK

chippery *n* a collection of devices or circuits built from silicon chips
□ see SILICON CHIP

choke *n* INDUCTOR

chromakey *n* a way of combining two video colour images in a TV
studio to create special effects (eg actors in a science fiction film
appearing to be on the surface of a planet). In this example, one
camera focusses on the actors who stand in front of a coloured
background, and a second camera focusses on a model of the
planet's surface. The background colour is removed electronically
from the first image and the two pictures are then combined
electronically to make the model appear to be the background.
□ see VIDEO CAMERA

chrominance *n* (the variation in) the amount of colour of an image
(eg a colour TV picture) – compare LUMINANCE □ see TELEVISION

chronograph *n* a watch or clock for accurately measuring time
intervals. A modern chronograph is electronic and uses a quartz
crystal and a specially designed integrated circuit to ensure accuracy
and reliability. Chronographs are especially useful in timing
competitive sports (eg athletics and sailing), and generally have a
digital readout though analogue readouts are by no means obsolete.
Some chronographs incorporate a sensor to provide a readout of

pulse rate, either by finger touch or by using a lightweight harness fitted to the chest. □ see DIGITAL WATCH

CIA – see COMMUNICATIONS INTERFACE ADAPTER

circuit *n* **1** an arrangement of conductors and components having a useful function <*a lighting* ~> □ see CIRCUIT BOARD **2** a path for two-way communication (eg between computer terminals) □ see CHANNEL

circuit board *n* a board (eg a printed circuit board) on which components are connected together to produce a circuit □ see PRINTED CIRCUIT BOARD

circuit breaker *n* a device operated by heat or magnetism which automatically turns off electrical power in the event of a short-circuit or overload □ see FUSE

clear *vb* to set to zero the binary data (usu a byte) in one or more locations in a computer memory

clipping *n* the removal, usu unintentionally, of the crests of the waveform of an analogue signal producing distortion of the signal <~ *in an audio amplifier*> □ see ZENER DIODE

CLK – see CLOCK

clock *n* (abbr **CLK**) **1** any circuit which provides a series of electrical ticks (or pulses) at a constant frequency. For example, a very stable crystal clock is used to synchronize the various functions of a microprocessor. □ see CRYSTAL CLOCK **2** the terminal of an integrated circuit (eg a microprocessor) to which pulses of constant frequency are applied

clock rate *n* the number of pulses per second produced by a special oscillator (the clock) in digital systems. In a computer a clock controls the speed at which data is passed from one part of the computer system to another. □ see CLOCK

closed *adj, of a switch* having two or more contacts which are touching each other so that current can flow in a circuit – compare OPEN □ see CLOSED CIRCUIT

closed-circuit *adj, of a TV installation* interconnected by cables within a local area

closed circuit *n* a circuit in which the components are connected so as to form a continuous path through which current can flow – compare OPEN CIRCUIT

closed-circuit TV *n* (abbr **CCTV**) – see CLOSED-CIRCUIT

closed-loop *adj, of a control system (eg a servosystem)* having a means of controlling the output by feeding a proportion of the output signal back to the input so as to oppose the input – compare OPEN-LOOP □ see NEGATIVE FEEDBACK

closed-loop gain *n* the voltage gain (amplification) of an amplifier when a proportion of its output signal is fed back to its input. The closed-loop gain of an operational amplifier is determined by the values of the resistors in the circuit which feeds back a proportion of the output voltage of the amplifier to its inverting input – compare

OPEN-LOOP GAIN □ see NEGATIVE FEEDBACK, OPERATIONAL AMPLIFIER

closed user group *n* (abbr **CUG**) a group of people who have exclusive access to a body of stored information. Viewdata offers various services (eg computer software) to closed user groups. □ see TELESOFTWARE

cluster *n* a group of components (eg microprocessors), terminals, or peripherals in close proximity to each other and under the control of a central device (eg a computer)

clutter *n* any unwanted shapes or patterns which appear on a display (eg a radar screen) and which make it difficult to see the image which should be there □ see INTERFERENCE

CMOS – see COMPLEMENTARY METAL-OXIDE SEMICONDUCTOR

CMOSL – see COMPLEMENTARY METAL-OXIDE SEMICONDUCTOR LOGIC

CNC machine *n* COMPUTER NUMERICALLY-CONTROLLED MACHINE

coaxial cable *n* (abbr **coax**) an electrical conductor (eg a TV aerial lead) for carrying high frequency signals. A coaxial cable has a central wire surrounded by a layer of insulation and then by a sheath of copper mesh (braid) which reduces electrical interference between neighbouring coaxial cables.

Cobol *n* [common business oriented language] a widely used high-level computer language which was specially devised with the intention of making it readily usable by nonmathematicians and which is of particular use for commercial applications □ see HOPPER

code *n* any system of characters (eg Morse code), or of conventions (eg the ASCII code) which represents information in a suitable form for transmission from one place to another □ see MACHINE CODE, BINARY CODE

code converter *n* an integrated circuit made from logic gates which changes digital information from one coded form to another (eg a BCD-to-decimal converter) □ see DECODER, ENCODER

coherent light *n* any light, esp the light emitted by a laser, which comprises many waves all in step (in phase) with each other □ see LASER

coil *n* a length of copper wire wound in one or more loops to form a spiral. The main purpose of a coil is to concentrate the magnetic effect of a current flowing through it. Coils have many different functions; a coil wound on a ferrite rod is used to improve the sensitivity of a radio receiver; a relay uses a coil wound on iron to produce a strong magnetic field to operate a lever to open and close switch contacts; transformers have two coils between which electrical power is transferred in alternating current circuits. □ see INDUCTOR, TRANSFORMER

cold boot *also* **cold start** *n* the way a computer automatically gets into its correct operating condition when it is first switched on, ie starting from 'cold'. An internal circuit sends a signal to the reset line of the microprocessor so that it runs a short program stored in ROM. This program checks all areas of RAM to find out how much memory is available to the user, and then jumps to the computer's operating system – compare WARM BOOT □ see ²BOOTSTRAP, OPERATING SYSTEM

cold start *n* COLD BOOT

collector *n* the one of the three terminals of a bipolar transistor towards which current flows □ see BIPOLAR TRANSISTOR

colour code *n* a set of coloured bands printed on some electronic components, esp resistors and capacitors, so that the value of the component can be easily ascertained □ see PREFERRED VALUES

colour television *n* a television receiver which makes use of the fact that almost any colour can be obtained by mixing the three primary colours of red, green, and blue in the correct proportions. A colour TV transmission contains three (chrominance) signals corresponding to these three primary colours. These three signals operate three electron guns which are designed to scan separately over red, green, and blue phosphor dots (in groups known as triads) on the screen of a cathode-ray tube. To make sure that the aim of each gun is precise and strikes the right coloured phosphor, a metal shadow mask pierced with about 400 thousand holes is placed close to the screen □ see TELEVISION, TELEVISION RECEIVER, ELECTRON GUN, PHOSPHOR, RGB GUNS, SHADOW MASK TUBE

Colpitts oscillator *n* a transistor circuit used for generating frequencies usu for radio transmitters. The Colpitts oscillator makes use of two capacitors which are part of a tuned circuit in the transistor's collector circuit – compare HARTLEY OSCILLATOR

column *n* a vertical arrangement of data – compare ROW

combinational logic *n* a digital logic circuit (eg a NAND gate) which makes decisions based on information it happens to be receiving at the moment, ie it does not have a memory. Combinational logic produces an output combination of logic 1's and 0's which is uniquely determined by a particular combination of 1's and 0's at two or more inputs – compare SEQUENTIAL LOGIC □ see GATE 1

command *n* a computer programming instruction (eg Print) telling a computer what to do

common emitter amplifier *n* an amplifier based on a single bipolar transistor connected so that both the input and output signals have the transistor's emitter terminal as a common reference. The common emitter amplifier is a general-purpose low-power amplifier used in devices of many different types, esp in radios, hi-fi systems and TVs. □ see BIPOLAR TRANSISTOR, COMMON SOURCE AMPLIFIER

common-mode rejection *n* the useful property of a differential amplifier (eg an operational amplifier) that enables it to amplify the difference in the signal strengths between its two inputs and to reject signals which are common to both inputs. Common-mode rejection allows instrumentation circuits (eg an electrocardiograph) to be designed to reject unwanted interference (eg mains hum).

common source amplifier *n* an amplifier which uses a single field-effect transistor connected so that both the input and output signals have the source terminal of the transistor as a common reference. The common source amplifier is widely used in hi-fi amplifiers and for many types of integrated circuit, since it has an extremely high input resistance, a low output resistance, and a voltage gain greater than 1. □ see FIELD-EFFECT TRANSISTOR, COMMON EMITTER AMPLIFIER

communicate *vb* to send information from one place to another – **communicator** *n*

communications interface *n* connectors and cable allowing a computer to be connected to a modem or other peripheral device (eg a printer) □ see RS-232 INTERFACE, IEEE-432 INTERFACE

communications interface adapter *n* (abbr **CIA**) an integrated circuit enabling communication between a computer and a modem or other peripheral □ see PERIPHERAL INTERFACE ADAPTER, UART

communications satellite *n* any of a large number of artificial satellites put in orbit to relay data, radio, and TV programmes between distant Earth-based receiving stations. Most communications satellites carry solar panels to power the complex electronic equipment carried aboard them. □ see DIRECT BROADCAST SATELLITE, INTELSAT, GEOSTATIONARY

communications system *n* an arrangement of components, circuits, and equipment which enables messages to be sent from one place to another □ see TELEPHONE, TELECOMMUNICATIONS, COMMUNICATIONS SATELLITE

compact disk *n* a plastic disk about 120 millimetres in diameter and 1.2 millimetres thick which stores audio information in digital form rather than in analogue form as on a conventional black vinyl disk. The digital information is stored in the form of microscopically small pits, each about one thousandth of a millimetre long and about a tenth as deep, arranged in a continuous spiral track so fine that fifty tracks are no wider than a human hair. Sixty minutes of recorded sound requires about ten million such pits. The disk is coated with a layer of reflecting aluminium over which is placed a protective film of transparent plastic. The compact disk is rotated at a precise speed in a player in which a finely focussed laser beam is reflected from the pits as a stream of 1s and 0s to be detected by a photodiode. The laser beam is mounted on a motor-driven servo-controlled carriageway which can track across the disk steadily during normal playing, or be moved rapidly to any required cue, or remain stationary in the 'pause' mode. Other servos monitor the focussing of the beam continuously and keep the beam centrally on the track. Synchronizing data within the pulse stream is also read and used to maintain the exact rotating speed of the disk. The laser beam focusses so accurately on the surface of the disk that only the pits are read not the dust or other irregularities on its surface. Although compact disks are at present more expensive than vinyl disks and tapes, the sound produced is almost free of distortion (eg surface hiss) and the system is compact and rugged enough for use in a car. □ see DIGITAL RECORDING, VIDEO DISK

comparator *n* also **discriminator** an analogue circuit based on an integrated circuit operational amplifier which compares the voltages of two input signals and produces an output when the signals are of different strengths. Comparators are extensively used in circuit design (eg in digital voltmeters and thermostats). □ see SCHMITT TRIGGER, MAGNITUDE COMPARATOR

compatible *adj, of an electronic system, esp a computer* capable of operating with the same hardware and/or software as another system – **compatibility** *n,* **compatibly** *adv*

compile *vb* to convert a program in a high-level language (eg Basic) into object code (eg machine code) □ see COMPILER, MACHINE CODE

compiler *n* a program which converts a source code program (eg Basic), which the programmer can readily understand, into a machine code or other object code program which the computer can understand □ see SOURCE CODE, MACHINE CODE, INTERPRETER

complementary *adj* mutually supplying each other's lack *<a ~ pair of transistors>* □ see COMPLEMENTARY PAIR

complementary metal-oxide semiconductor *n* (abbr **CMOS**) a switching device comprising an n-channel and a p-channel field-effect transistor connected in series with each other on the same piece of silicon. A wide range of digital integrated circuits are made from complementary metal-oxide semiconductors. They are characterized by having a low power consumption, making them useful in watches and other low-power battery-operated devices. Their main disadvantage is that they are liable to damage caused by static electricity. □ see N-CHANNEL, P-CHANNEL, METAL-OXIDE SEMICONDUCTOR FIELD-EFFECT TRANSISTOR, ELECTROSTATIC DISCHARGE

complementary metal-oxide semiconductor logic *n* (abbr **CMOSL**) a common type of logic gate in integrated circuit form characterized by medium-speed operation, very low power dissipation, and a wide operating supply voltage range (+3 volts to 18 volts). The digital circuits in watches, calculators, and other portable equipment operated from batteries are usu based on CMOSL – compare TRANSISTOR-TRANSISTOR LOGIC □ see COMPLEMENTARY METAL-OXIDE SEMICONDUCTOR

complementary pair *n* one npn and one pnp bipolar transistor, or one n-channel and one p- channel CMOS transistor, coupled together in a circuit (eg an audio amplifier) so that the amplification of an alternating signal is shared efficiently between them □ see

PUSH-PULL AMPLIFIER

complex *adj* difficult to work out or to see in detail <*a ~ integrated circuit*> – **complexly** *adv*, **complexity** *n*

component *n* any part of electronic hardware having a particular purpose <*resistors, capacitors, screws, and integrated circuits are examples of ~s making up an amplifier*> □ see BUILDING BLOCK

compunications *n* a word coined to describe the coming together of computers and communications for some useful purpose <*Viewdata is an example of ~*>

compute *vb* to determine or calculate using a calculator or computer <*to ~ the path of a satellite*> – **computable** *adj*, **computability** *n*, **computation** *n*

computer *n* a programmable electronic device designed for storing, retrieving, and processing data □ see MICROCOMPUTER, MICROPROCESSOR, COMPUTER GENERATIONS, SOFTWARE, FIFTH GENERATION COMPUTER

computer-aided design *n* (abbr **CAD**) a computer system for helping engineers, architects, scientists, car manufacturers, town planners, interior designers, and others to design their products and services. Computer-aided design involves the use of a keyboard, light pen, and other aids coupled to a computer so that information drawn on a VDU can be modified. When linked with computer-aided manufacturing (CAM) the facility is known as CAD/CAM – compare COMPUTER-AIDED MANUFACTURING □ see CAD/CAM, GRAPHICS TERMINAL, LIGHT PEN

computer-aided instruction *n* (abbr **CAI**) the use of the computer to enhance traditional teaching methods. Computer-aided instruction provides instruction through the use of graphics, animation, or sound to make material more interesting than it would be in a textbook or on a blackboard. It allows a student to progress at his/her own pace while keeping a check on performance. When linked with computer-aided learning (CAL), the facility is known as CAI/CAL – compare COMPUTER-AIDED LEARNING

computer-aided learning *n* (abbr **CAL**) the use of the computer to enhance the process of learning. Computer-aided learning provides an environment in which the student can explore and learn, which results in a much better understanding than instruction alone. When linked with computer-aided instruction (CAI), the facility is known as CAI/CAL – compare COMPUTER-AIDED INSTRUCTION □ see LOGO, EXPERT SYSTEM

computer-aided manufacturing *n* (abbr **CAM**) the use of computers to control how something is made. CAM includes controlling the actions of robots on a production line and helping to make manufacturing decisions. When linked with computer-aided design (CAD) the facility is known as CAD/CAM – compare COMPUTER-AIDED DESIGN □ see CAD/CAM, COMPUTER-NUMERICALLY CONTROLLED MACHINE

computer chess *n* a dedicated microprocessor-based unit, or a program which can be loaded into a computer, that enables a person to play a game of chess with the computer

computer crime *n* the tampering with a computer system by unauthorized people for information or financial gain. For example, a person might attempt to use a microcomputer to transfer funds from one account into his or her own or a fictitious account. Computer crime has increased with the greater use of computers so special security devices are built into computer programs to prevent this. □ see ELECTRONIC FUNDS TRANSFER, DONGLE

computer engineer *n* one who designs or maintains computer hardware – compare SOFTWARE ENGINEER

computer generations *n* the 'family history' of computers in terms of the main technological advances used in their development. The first generation was based on the thermionic valve and is regarded as having been 'born' in the early 1950s. Second generation computers came along in the late 1950s and early 1960s and used individual transistors. During the 1960s integrated circuits containing up to a few hundred transistors were used in third generation computers. Integrated circuits made from several thousand transistors were used in the fourth generation computers of the 1970s. And in the 1980s, fifth generation computers based on very large-scale integrated circuits containing more than a million transistors on a single chip are being developed. □ see ENIAC, MOORE'S LAW, VERY LARGE SCALE INTEGRATION, FIFTH GENERATION COMPUTER

computer graphics *n* the processing and generation of visual information using a computer linked to a keyboard, VDU, and other peripherals □ see GRAPHICS TERMINAL, COMPUTER-AIDED DESIGN

computerize *vb* to make use of or equip with computers <*a* ~d *factory*> – **computerization** *n*

computerized axial tomography *n* (abbr **CAT**) a diagnostic tool for producing detailed images of the body's internal organs using X-rays and computer processing. CAT uses a beam of X-rays to scan the body in a series of thin slices so that shadows of tissue and organs in front of and behind the part being scanned do not show up on the screen. A sensitivity 50 times higher than that of ordinary X-ray techniques is obtained. □ see ULTRASONIC SCANNER

computer literacy *n* the ready acceptance and use of computers in everyday life. Computer literacy includes an understanding of the way a microcomputer works and how to program it, the commercial applications of computers, and the social effects of the widespread use of computers in society. □ see INFORMATION TECHNOLOGY

computernik *n* an enthusiastic computer hobbyist who spends a great deal of time with computers □ see CHIP HEAD, HACK

computer numerically-controlled machine *also* **CNC machine** *n* a machine tool (eg a lathe) which is operated under computer control □ see COMPUTER-AIDED MANUFACTURING

computer program *n* – see PROGRAM

computer science *n* the study of how computers work, the mathematics of their operation, and the languages with which we communicate with them □ see LANGUAGE, BINARY ARITHMETIC, KNOWLEDGE ENGINEERING, FIFTH GENERATION COMPUTER

computer system *n* one or more (micro)computers and associated peripherals *<a home ~>*

computer users association *n* (abbr **CUA**) USER GROUP

computing *n* **1** the process of working with numbers to solve a problem *<navigational ~>* □ see COMPUTE **2** the science, activity, or profession of using numbers *<a course on ~>* □ see COMPUTER SCIENCE

concatenate *vb* to add or link together *<to ~ two strings in a computer program>* – **concatenation** *n*

concertina fold paper *n* FANFOLD PAPER

concurrent *adj* happening at the same time *<~ processing>* – **concurrent** *n,* **concurrently** *adv* □ see PARALLEL

conditional branch *n* a jump from one point to another in a program which will only be carried out by the computer if certain conditions (eg the presence of a 0 in a register) are met – compare UNCONDITIONAL BRANCH

conductance *n* – see SIEMEN

conduction *n* the flow of electricity or heat through a substance □ see CONDUCTIVITY

conduction band *n* a band of atomic energy levels occupied by electrons which are loosely bound to the parent atom and which are responsible for electrical conduction in a material. No electrons can occupy the conduction band in an electrical insulator such as glass. Many electrons occupy the conduction band in an electrical conductor such as copper. In semiconductors, such as silicon and germanium, electrons can be made to move into the conduction band by heating the semiconductor or by adding small amounts of impurities. The latter technique is the basis of producing useful p-type and n-type semiconductors from silicon for use in transistors and integrated circuits. □ see SEMICONDUCTOR, VALENCY, ENERGY GAP, P-TYPE, N-TYPE

conductive elastic wristband *n* a closed elasticated wristband which is fitted over the wrist and connected to earth to discharge any static electricity which might accumulate on a person's body. It is intended to protect components (eg CMOS devices) which a person might be handling from damage by static electricity. □ see STATIC ELECTRICITY, COMPLEMENTARY METAL-OXIDE SEMICONDUCTOR

conductivity *n* a measure of how well electricity flows through a substance *<copper has good ~, pure silicon poor ~>* – compare RESISTIVITY

conductor *n* a material (eg copper) or a path (eg a track on a circuit board) through which electricity flows (easily) □ see CONDUCTIVITY

configuration *n* the way parts of a circuit or system are arranged <*the ~ of a computer system*> – **configure** *vb* □ see ARCHITECTURE
connect *vb* to make electrical contact between two devices or circuits so that information can flow between them <*to ~ a computer to a printer*> – **connection** *n*
connector *n* any device which enables information to flow between two or more circuits or systems <*a printed circuit board ~*> □ see FIBRE OPTICS CONNECTOR, BNC
console *n* an array of controls and indicators for controlling and monitoring the working of something <*an instrument ~ on the Concorde's flight deck*>
constant *n* a number (eg pi) with a value that remains unchanged in calculations – compare ¹VARIABLE
consumer device *n* an electronic component which is manufactured mainly for use in equipment used by the general public □ see ELECTRONIC CAMERA, KITCHEN ELECTRONICS
consumer electronics *n* electronics equipment designed for home use. Once upon a time, the only electronic device to be found in the home was a radio. Now consumer electronics includes the record player, television, video recorder, electronically controlled toaster, washing machine, and a whole host of other gadgets to entertain, inspire thought, and increase home comfort. The home computer is the latest device to have a real impact on home life. □ see HOME COMPUTER
contact *n* the part of a switch through which current flows when it meets a similar part <*relay ~s*>
¹**control** *n* a device for directing or regulating something <*a joystick ~*>
²**control** *vb* to direct or regulate something <*to ~ a robot arm*> – **controllable** *adj*, **controllability** *adv*
control bus *n* a group of tracks on a printed circuit board of a microcomputer which carries control signals in binary code from the microprocessor to memory and other chips
controller *n* any device (eg a drill speed controller) which automatically or manually controls the operation of something □ see DIMMER, THERMOSTAT
control lines *n* conductors (eg wires or copper tracks on printed circuit boards) in an electronic system which carry signals for particular operations □ see CONTROL BUS
control panel *n* an arrangement of switches and lights used to monitor and control an operation <*a ~ in a power station*>
control processor *n* a microprocessor for controlling the operation of other microprocessors in a computer system
control program *n* **1** a program that controls the operations of other programs **2** a program stored in a read-only memory for controlling a device <*the ~ of a washing machine*>
control station *n* a usu complex console or building which

masterminds the operation of something <*a ~ for monitoring Shuttle flights*> □ see EARTH STATION

convention *n* any standard and accepted procedure or set of symbols, abbreviations, etc together with their meanings □ see CONVENTIONAL CURRENT

conventional current *n* a generally accepted rule that electricity flows from the positive to the negative potential in a circuit. Although in most solid conductors, electricity is a flow of electrons from the negative to the positive potential, the rule does not usu make circuit analysis difficult.

convergence *n* the coming together of computers, telecommunications, and office automation to produce the electronic office □ see ELECTRONIC OFFICE, INFORMATION TECHNOLOGY

converter *n* a device for changing electrical signals from one form to another <*an AC-to-DC ~*> □ see ANALOGUE-TO-DIGITAL CONVERTER

coordinates *n* two numbers which identify the horizontal and vertical positions of a pixel or graphics character on the screen of a VDU □ see PLOT

copy *vb* **1** to make a duplicate of data <*to ~ a program on disk to magnetic tape*> **2** to read data in one location in a storage device and put it in another location while leaving the original data unchanged – **copy** *n*

core memory *n* an obsolete random access computer memory device consisting of many magnetic rings (tori) threaded onto a grid of interconnecting wires. Electric current flowing through a wire magnetizes the rings and the two possible directions of magnetization represent the binary states 1 and 0.

corner frequency *n* a characteristic region on the graph of voltage gain versus frequency for an amplifier, where the gain is 3 decibels below maximum gain. Corner frequency is used to assess the performance of the amplifier at high frequencies. □ see BANDWIDTH, DECIBEL, FILTER

corrupt *vb* to cause errors or faults in something <*to ~ a program on magnetic tape*> – **corruption** *n* □ see BUG, GLITCH

coulomb *n* the unit for measuring electrical charge. A current of 1 ampere through an electrical conductor is a rate of flow of electricity of 1 coulomb per second □ see ELECTRON, PROTON

counter *n* any of a number of integrated circuits (eg a BCD counter) consisting of a set of flip-flops for counting pulses entering the input. The count is stored as a binary number □ see REGISTER

coupling capacitor *n* also **blocking capacitor** a capacitor used to pass a varying signal from one part of a circuit to another (eg between two stages of amplification in an amplifier) while isolating the differing steady signals in the two circuits – compare DECOUPLING CAPACITOR □ see CAPACITOR

CPA – see CRITICAL PATH ANALYSIS

CP/M *trademark* [control *p*rogram for *m*icrocomputers] a complex
operating system which is being made available to an increasing
number of microcomputers so that they can use programs written for
other machines which also have this operating system. CP/M is the
trademark of Digital Research. A very large amount of software is
available to the user of a CP/M.

CPS 1 – see CHARACTERS PER SECOND **2** – see CYCLES PER SECOND
CPU – see CENTRAL PROCESSOR UNIT

crash *vb* to become suddenly incapable of working properly <*the
program* ~ed> – **crash** *n*

crimp connector *n* a small connector which is joined to the end of a
wire by squeezing it to grip the wire tightly

critical angle *n* the angle at which light must be reflected off the
inside wall of an optical fibre in order for the light to be transmitted
along the fibre without coming out of its sides □ see OPTICAL FIBRE,
OPTICAL COMMUNICATIONS

critical path analysis *n* (abbr **CPA**) a (computer-assisted) technique
for optimizing manufacturing and construction processes. A
shipyard, for example, would use critical path analysis to plan the
allocation of men and materials in the right order so that an estimate
of the delivery date for a completed ship can be made.

CRO – see CATHODE-RAY OSCILLOSCOPE

crocodile clip *n* a spring-operated metal connector attached to a
wire and having two toothed jaws for gripping other wires and
terminals

crossover distortion *n* a particularly unpleasant sounding distortion
of a signal in an audio system caused by incorrect operation of the
complementary pair of transistors in the push-pull output stage of
the amplifier □ see PUSH-PULL AMPLIFIER

crosstalk *n* unwanted signals on a communications channel which
come from another channel. Hi-fi amplifiers and telephone systems
can be affected by crosstalk. □ see INTERFERENCE

crowbar protector *n* an integrated circuit intended to protect an
electronic circuit against high voltage transients on the power supply
line which could damage some components in the circuit. A crowbar
protector is wired across the power supply line and detects when the
supply line voltage rises above a certain level. It then triggers a
thyristor which short-circuits the power supply, causing a fuse to
open or in some other way isolating the circuit from the power
supply. □ see THYRISTOR, TRANSIENT

CRT – see CATHODE-RAY TUBE

cruise missile *n* a self-propelled low-flying atomic bomb which is
programmed to reach a target automatically. A cruise missile carries
an on-board computer which compares the view of the ground over
which it flies with an image of the target area stored in its memory.

cryogenic *adj* involving the study or use of very low temperatures
close to absolute zero (ie 0 to 5K) <~ *memory*> – **cryogenics** *n*

□ see ABSOLUTE TEMPERATURE SCALE, SUPERCONDUCTOR, CRYOGENIC MEMORY, JOSEPHSON JUNCTION, IRAS

cryogenic memory *n* any memory device (eg Josephson memory) which operates at near absolute zero (about −273C) to take advantage of the low resistance of conductors at this temperature □ see JOSEPHSON JUNCTION

cryptography *n* the preparation of secret codes, usu by electronic means, for security purposes □ see DONGLE, SECURITY ELECTRONICS

¹crystal *n* **1** a material comprising atoms arranged in an orderly way. A crystal usu reflects this orderly arrangement by having a geometrical shape with flat faces. **2** a quartz or lead zirconium titanate crystal which is used to make very stable oscillators <*a ~ oscillator in a computer*> □ see CRYSTAL CLOCK, DIGITAL WATCH, MICROPROCESSOR, PIEZOELECTRICITY

²crystal *adj* of devices (eg a crystal microphone) that make use of the piezoelectric properties of quartz and certain ceramics □ see CRYSTAL CLOCK

crystal clock *also* **crystal oscillator** *n* a high frequency (eg 4 megahertz) oscillator controlled by the vibrations of a crystal. A crystal clock is used for controlling and synchronizing the various operations performed by a microcomputer. Binary dividers reduce the high frequency to lower frequencies to control functions such as the rate at which data is transferred to and from tape or disk. □ see CRYSTAL, FREQUENCY

crystal oscillator *n* CRYSTAL CLOCK

CUG – see CLOSED USER GROUP

current *n* the flow of electricity along an electrical conductor from a point of high to lower electrical potential □ see AMPERE, POTENTIAL, CONVENTIONAL CURRENT

current differencing amplifier *n* (abbr **CDA**) a type of integrated circuit operational amplifier (op amp) which produces an output voltage proportional to the difference in current flowing into its two inputs. This function differs from the normal op amp which operates by sensing a voltage difference at its two inputs. The advantage of the current differencing amplifier is that it is cheaper than the normal op amp, and it has a higher bandwidth making it suitable for audio amplifiers. □ see OPERATIONAL AMPLIFIER

current hogging *n* the drawing (hogging) of more current by either of a pair of transistors which are supposed to take an equal share of current. Current hogging makes a circuit perform badly, although careful circuit design can minimize its effect. □ see THERMAL RUNAWAY

cursor *n* a movable marker (eg a flashing square or arrowhead) on the screen of a VDU which indicates where the next character will be printed □ see TURTLE, BLINK

cursor control keys *n* keys for moving a character round the screen of a VDU using the keyboard arrows indicating up, down, right, and left

custom design *n* the design and manufacture of a device (usu an integrated circuit) by a component manufacturer to suit a particular customer's needs

cutoff frequency *n* the maximum or minimum frequency passed by a filter □ see FILTER, BANDWIDTH

CW – see CARRIER WAVE

cybernetics *n* the study of how communications and control systems can be combined to simulate human functions. Cybernetics increasingly makes use of the latest achievements of microelectronics, particularly computing, pattern recognition, and voice recognition. – **cybernetic** *adj* □ see ARTIFICIAL INTELLIGENCE, ROBOTICS, ANDROID, CYBORG

cyborg *n* [*cyb*ernetic *org*anism] a living organism which uses mechanical and/or electronic parts. A person with a heart pacemaker is a cyborg. □ see CYBERNETICS, ANDROID

cycle *n* a complete sequence of a wave pattern (eg a sound wave) which is repeated at regular intervals □ see FREQUENCY

cycles per second *n* (abbr **CPS**) the speed at which periodic signals take place <*mains alternating current is produced at 50 ~*> □ see HERTZ

cycle time *n* the time taken for a microprocessor or other data processing device to carry out one complete set of operations □ *see* MICROPROCESSOR

cyclic storage *n* computer memory in which data moves in relation to where it is read □ *see* MAGNETIC DISK, MAGNETIC BUBBLE MEMORY, CHARGE-COUPLED DEVICE

D

D the symbol used in the hexadecimal system for the decimal
number 13 (eg the hexadecimal number 2D equals the decimal
number 2 × 16 + 13 = 45) □ see HEXADECIMAL

DAC – see DIGITAL-TO-ANALOGUE CONVERTER

daisy chain *n* a number of devices (eg disk drives) connected in
series (ie one after the other) – **daisy chain** *vb*

daisywheel *n* – see DAISYWHEEL PRINTER

daisywheel printer *n* a printer producing high quality hard copy (eg
of program listings) on paper. The daisywheel printer has a
printwheel with characters ('petals') round its edge which are moved
in front of a hammer to be struck against the paper – compare DOT
MATRIX PRINTER

damping *n* the reduction or elimination of unwanted oscillation in
an electrical device or circuit *<~ of a control system>*

darlington pair *n* a connection of two bipolar transistors so that
their combined current gain is equal (theoretically) to the product of
their gains. The darlington pair requires such a small current to
switch it on that it is often used to couple a low-power device such as
a microprocessor to a high-power device such as a motor. For
interfacing computers to the real world, darlington pairs are
available in multiples in integrated circuit packages. □ see BIPOLAR
TRANSISTOR, GAIN, BUFFER

data *n* any information (eg facts, statistics, measurements, and
symbols) which can be processed or communicated *<computer ~>*
□ see INFORMATION TECHNOLOGY

data acquisition *n* DATA CAPTURE

databank *n* a collection of data stored in a computer system

database *n* a collection of data that is stored in a computer system
and consists of indexed references (eg subject headings, keywords
and phrases, or literature titles) to information which a user can
quickly gain access to using a keyboard. A home computer linked via
the telephone to a central computer can be used to obtain
information from a database. □ see PRESTEL, EXPERT SYSTEM

data bus *n* an electrical pathway between the microprocessor and
memory in a computer that consists of a set of wires (usu tracks on a
printed circuit board) along which binary signals flow one byte at a
time that represent information to or from a piece of external
hardware (eg a printer) □ see ADDRESS BUS, MEMORY

data capture *also* **data acquisition** *n* any method for collecting data
and converting it into a form which a computer can use

data latch *n* LATCH

data logger *n* a device for recording data from a number of sources
simultaneously for later use. Thus a data logger might be used to
record (ie to log) data from a remote weather station over a period

of several weeks, and to send this data via radio to a central
receiving station.

data processing *n* also **information processing** the acquisition,
recording, and manipulation of data by electronic, esp computer-
based, means

data retrieval *n* the selection and extraction of data from a
database, usu through a keyboard connected to a computer □ see
DATABASE, VIEWDATA

data sink *n* the end of a communications channel at which data is
received – compare DATA SOURCE

data source *n* the end of a communications channel from which
data is transmitted – compare DATA SINK

daughterboard *n* – see MOTHERBOARD

dB – see DECIBEL

DBS – see DIRECT BROADCAST SATELLITE

DC – see DIRECT CURRENT

deaf aids *n* electronic devices for helping the deaf lead a more
normal life. The hearing aid which fits into the ear is the best known
deaf aid and one which has benefitted from the use of miniature
microelectronic devices. Young people deaf from birth can be taught
to speak using computers which display the shape of the mouth and
the voice patterns of particular sounds. A long-term aim is to
develop electronic devices that will replace damaged inner ears by
stimulating the auditory nerves directly. □ see BLIND AIDS

debouncer *n* a circuit, usu based on a Schmitt trigger, which
removes unwanted secondary signals produced by contact bounce in
a mechanical switch. A debouncer ensures that a digital circuit
responds to just one signal for each event fed to it (eg the operation
of a switch on a production line). □ see SCHMITT TRIGGER

debug *vb* to correct any errors in a computer program or in the
circuits of the computer itself

decade *adj* counting or processing in units of ten <*a ~ counter* > –
decade *n*

decade counter *n* also **scale-of-ten counter** a binary counter which
has ten outputs each of which is activated in turn when the counter
receives a sequence of pulses at its input. Decade counters are used
for directing a sequence of operations in communications and control
systems. □ see BINARY COUNTER

decay *n* the fall in strength of something <*the ~ of charge on a
capacitor*> – **decay** *vb*

decibel *n* (symbol **dB**) a unit used for comparing the strengths of
two signals. The decibel is often used to measure the intensity of
sound relative to the weakest sound that the ear can detect (eg a
sound intensity of 120dB causes pain). The unit is also used for
measuring the voltage gain of an amplifier (eg an amplifier which
amplifies an input signal a thousand times has a gain of 60dB). Since
the decibel is defined as 20 times the logarithm of the ratio of the

strengths of two signals, a further increase of one thousand times is a doubling of the gain measured in decibels, ie to 120dB.

decimal *adj* of or being a number system having 10 digits – compare BINARY

decoder *n* any device which converts information from a coded form (eg binary code) to another form (eg decimal) which is more readily understood either by a person or an electronic system – **decode** *vb*

decoupling capacitor *n* a capacitor used to remove unwanted varying signals (eg power supply ripple) at some point in a circuit – compare COUPLING CAPACITOR

decrement *vb* to reduce the value of a variable usu by one (eg in Basic LET A=A−1), or the value of the contents of a register or memory location – compare INCREMENT – **decrement** *n*

dedicated *adj, of an electronic device (eg a microprocessor)* programmed to do one particular job (eg controlling a washing machine) – **dedicate** *vb*

default *n* something assumed or done by a computer system if no alternative is specified

De Forest, Lee. American inventor of the triode valve in 1903. □ see TRIODE

delay *n* the time taken between the start and finish of something. Delay is unavoidable (and sometimes desirable) in electronic systems. For example, the time taken for binary signals to flow through logic gates and other parts of a computer circuit restricts the speed at which computer systems operate, and the slight delay caused by radio waves travelling to and from communications satellites is slightly annoying to users making telephone calls across the world. However, in some electronic systems it is necessary to create a delay electronically to ensure that various signals synchronize and flow smoothly through a system. □ see DELAY LOOP, MONOSTABLE, GALLIUM ARSENIDE

delay loop *n* a short program in Basic which uses the instructions FOR, TO, and NEXT to provide a time delay in the execution of a program

demodulate *vb* to recover information (eg music) from a carrier wave (eg a radio wave) – compare MODULATE – **demodulation** *n* □ see AMPLITUDE MODULATION, PULSE CODE MODULATION

demultiplexing *n* a method of recovering the individual messages from a single communications channel that carries several different messages simultaneously <*the telephone uses* ~> – compare MULTIPLEXING – **demultiplexing** *n,* **demultiplex** *vb*

density *n* – see PACKING DENSITY

depletion region *n* also **space charge** the region across a reverse-biased pn junction in which there are only a few electrons and holes thus making the region a poor conductor of electricity. Rectifier diodes and bipolar transistors depend for their operation on the existence of the depletion region. □ see REVERSE BIAS, PN

JUNCTION, BIPOLAR TRANSISTOR

deposition *n* the process of laying down a thin layer of aluminium, silicon dioxide, or semiconductor impurity during the manufacture of an integrated circuit on the surface of a silicon chip □ see METALLIZATION

desolder braid *n* copper wire in the form of a web which is used to remove excess solder from a circuit board. The wire is laid on the solder to be removed and heated with a soldering iron so that the melted solder clings to the wire and can be easily removed.

destination *n* a place (eg a register) to which data is sent – compare SOURCE

detector *n* a device used in the demodulation stage of an AM or FM receiver for recovering the original signal from the modulated carrier wave – **detection** *n,* **detect** *vb* □ see DEMODULATE

deterministic *adj* of or being a process or model (eg a computer simulation of traffic flow at an airport) controlled by force of circumstances – compare STOCHASTIC

DFS *n* [*d*isk *f*iling *s*ystem] a collection of files (a computer 'filing cabinet') on one or more magnetic disks

diac *n* a semiconductor device used in AC power control circuits (eg a drill speed controller) for switching on a triac effectively when the voltage across the diac reaches a particular voltage □ see TRIAC

dialects *n* variations in the syntax of a particular computer language □ see LANGUAGE

dibit *n* a piece of binary data consisting of two bits □ see BINARY DIGIT

dielectric *n* the insulating material (eg mica, glass, or polystyrene) between the two conducting plates of a capacitor □ see CAPACITOR, INSULATOR

differential amplifier *n* an amplifier (eg an operational amplifier) which is designed to respond to the difference in the two currents or voltages present at its two inputs □ see OPERATIONAL AMPLIFIER, COMMON-MODE REJECTION

diffusion *n* **1** the movement in a semiconductor (eg a transistor) of electrons or holes from a region of high concentration to a region of low concentration □ see BIPOLAR TRANSISTOR **2** the process of making p-type and n-type semiconducting regions in a chip of pure silicon by allowing selected impurities in gaseous form to enter the chip through precisely defined 'windows' in the chip □ see GASEOUS DIFFUSION, PHOTOLITHOGRAPHY, PHOTOMASK

digit *n* any of the ten numbers 0 to 9. The binary number system uses the digits 1 and 0. □ see DECIMAL, BINARY, HEXADECIMAL

digital *adj* **1** of or being a circuit or system that responds to or produces electrical signals of two values, logic high (binary 1) and logic low (binary 0) **2** *of a display (eg a watch face)* showing a quantity as a set of numbers – compare ANALOGUE

digital computer *n* a device which uses digital circuits (eg gates and counters) to process data. Digital computers have largely replaced analogue computers which work on continuously variable quantities – compare ANALOGUE COMPUTER

digital display *n* a readout of the value of something (eg the winning time of a sprint race) as numbers on a digital display (eg a seven-segment display) – compare ANALOGUE DISPLAY □ see DIGITAL MULTIMETER

digital frequency meter *n* an electronic instrument which measures the frequency of a waveform (eg a radio carrier wave) and displays the reading as a seven-segment digital display □ see PRESCALER

digital multimeter *n* a meter which uses digital circuits (eg counters) to measure and display, as numbers, the values of the electrical quantities volts, amperes, and ohms – compare ANALOGUE MULTIMETER

digital recording *n* a method of producing high quality sound recordings on vinyl disks by recording the original sound in digital form, rather than in analogue form, on a master tape. When making an analogue recording on tape (as in an ordinary cassette recorder) it is easy for the tape to pick up electrical noise from nearby machinery and this noise, plus a certain amount of tape 'hiss', is heard on playback. A digital recording does not suffer from these problems. In a digital recording, the amplitudes of the original sound patterns from microphones are sampled at about 50 000 times per second and each sample is converted into a binary number which is then stored on the master tape. Even though this tape may subsequently pick up electrical noise, any copies made from it only reproduce the digital signals. Thus any copy can be used to produce a master disk from which near perfect recordings on vinyl disks can be made. These disks are called digital recordings because of the original digital recording on the master tape; the information on them is still in analogue form however. The digital information on the master tape can easily be transferred to a compact disk which does hold its information in digital form. □ see COMPACT DISK, VIDEO DISK

digital-to-analogue converter *n* (abbr **DAC**) a device which changes a digitally coded signal (eg an 8-bit word) into an analogue signal of equivalent value. A digital-to-analogue converter enables a digital system (eg a microcomputer) to control the speed of a motor or the brightness of a lamp smoothly – compare ANALOGUE-TO-DIGITAL CONVERTER

digital tracer *n* DIGITIZER 1

digital watch *n* a small timepiece, usu worn on the wrist, which makes use of microelectronics to keep accurate time and to provide other useful functions. Most of the functions of a digital watch are under the control of a single integrated circuit controlled by the precise 'ticks' of a crystal oscillator. The energy which drives the circuit comes from a small mercury or lithium battery. A liquid crystal display is the preferred method for showing the time and this might also show 'hands' to simulate the face of an old-style mechanical watch. Push buttons are used to select time, date, and alarm functions, and any other facilities (eg games and calculator functions) the device might offer. A digital watch which also functions as a TV receiver has been built, and some can be programmed using radio signals to act as a portable database. In the long-term we shall probably use a digital watch to see and talk to another person anywhere in the world. □ see WRISTWATCH TV

digitize *vb* to convert information from analogue form to digital form □ see DIGITIZER, DIGITAL RECORDING, ANALOGUE-TO-DIGITAL CONVERTER

digitizer *n* **1** *also* **digital tracer, graphics tablet** a computer peripheral enabling diagrams to be 'copied' on the screen of a VDU. One type of digitizer consists of an electronic pen connected to the computer which is used to trace round a diagram placed on a special pad (a bit pad) which is also connected to the computer. A second type of digitizer makes use of an arm connected to one or more potentiometers which carries a stylus for tracing round a picture. □ see MOUSE, TRACK BALL ROLLER **2** ANALOGUE-TO-DIGITAL CONVERTER

DIL – see DUAL-IN-LINE

DIL switch *n* a group of usu four or eight miniature on/off switches contained in a dual-in-line package. A DIL switch is intended to be soldered in a printed circuit board where its small size saves space. The switches may have a slide or rocker action, and they are generally operated using a small pointed object (eg a pencil). □ see DUAL-IN-LINE

dimmer *n* a device (eg a wall-mounted unit) which has a knob or a touch-plate for controlling the brightness of a lamp

DIN *n* [*D*eutsche *I*ndustrie *N*orm] a German standards authority which specifies how connectors and similar devices should be wired up to audio hardware (eg cassette recorders) □ see DIN PLUG

DIN plug *n* a cylindrical connector having pins surrounded by a metal collar, which is plugged into a DIN socket so that electrical equipment (eg a computer and a printer or cassette recorder) can be connected together □ see DIN, D-TYPE CONNECTOR

diode *also* **diode valve** *n* a device which allows current to flow through it in one direction only. The thermionic diode has now largely been replaced by the semiconductor diode. Diodes are used in power supplies to convert an alternating current to a direct current. □ see RECTIFIER, PN JUNCTION, DETECTOR

DIP [*d*ual-*i*n-line *p*ackage] – see DUAL-IN-LINE

dipole aerial *n* also **Yagi aerial** a radio transmitting or receiving aerial (eg a roof-top TV aerial) consisting of one or more pairs of vertical or horizontal metal rods. The lengths of the rods are shorter for higher frequency radio transmissions – compare DISH AERIAL

direct access *n* a fast method of finding data in a storage medium, esp on magnetic disk or in random-access memory, where the access time is independent of the location of the data □ see RANDOM-ACCESS MEMORY

direct addressing *n* also **absolute addressing** the use of machine code to tell a microprocessor where to find the next piece of data, by letting the operand be the address where the data is stored □ see OPERAND, IMMEDIATE ADDRESSING, INDIRECT ADDRESSING

direct broadcast satellite *n* (abbr **DBS**) an artificial satellite in Earth orbit which broadcasts TV programmes direct to homes. Direct broadcast satellites are already in use in North America and provide people in remote areas with a TV service. A similar service is being considered for the UK during the 1980s using Olympus satellites built by British Aerospace. Householders and others who want to receive programmes from space would need to point a small rooftop mounted dish-shaped aerial to the point in the sky where the satellite 'hangs' in geostationary orbit. Transmissions from direct broadcast satellites will be scrambled so that only those subscribers who have the right decoding equipment connected to their TV sets will be able to receive programmes. The TV pictures do not, of course, originate in the satellite but are first beamed up to it from a ground station and then rebroadcast, after amplification, to selected areas of the Earth's surface. □ see COMMUNICATIONS SATELLITE, CABLE TV, FOOTPRINT, TRANSPONDER

direct coupler *n* MODEM

direct current *n* (abbr **DC**) a flow of electricity in one direction only $<\sim$ *flow from a dry battery through a lamp*$>$

directional *adj* of or being information flow (eg data on a computer bus) in one direction only – compare BIDIRECTIONAL

direct memory access *n* (abbr **DMA**) a facility that some computers have for allowing data to be transferred direct into memory without using a program under the control of the microprocessor. Direct memory access is of special value in high-speed memory systems such as disk storage where the data flow rate easily exceeds the capacity of a program to keep up with the data flow.

directory *n* a table on a disk filing system which gives information about the names, addresses, and sizes of files

disable *vb* to make a device (eg an integrated circuit counter) incapable of responding to data being fed to it. Many types of digital integrated circuits (eg microprocessors) have one or more terminals for the purpose of momentarily disabling a function – compare ENABLE

disassembler *n* a program which translates machine code into assembly language – compare ASSEMBLER – **disassemble** *vb*

discharge *n* **1** a usu brief flow of electricity $<a \sim of\ lightning>$ **2** loss of power from a cell or battery with age or use **3** loss of electrical charge stored in a capacitor – **discharge** *vb*

disconnect *vb* to remove a device from its source of power or information

discrete *adj, of an electronic component* being a single device or having a single purpose <*a transistor is a ~ semiconductor device*> – compare INTEGRATED

discriminator *n* COMPARATOR

dish aerial *n* a device with a dish-shaped collector for focussing radio signals onto an aerial in the centre of the dish. Large dish aerials are used for communicating with artificial satellites and interplanetary space probes since a large dish can collect a lot of weak signals. Similarly, radio astronomers use large dish aerials to examine radio emissions from distant galaxies. Ground-based microwave communications links also use dish aerials mounted on masts to send information in straight lines from one mast to another. □ see MICROWAVES, RADIO ASTRONOMY

disk *n* – see FLOPPY DISK, HARD DISK, WINCHESTER DISK, VIDEO DISK

disk drive *n* a program storage device for transferring data to and from magnetic disks. Fast and accurate reading and writing of data on the disk enables large amounts of data to be accessed and loaded quickly. □ see MAGNETIC DISK, DISK OPERATING SYSTEM

diskette *n* FLOPPY DISK

disk operating system *n* (abbr **DOS**) a program or a group of programs which instructs a computer how to read and store information on a magnetic disk. Some microcomputers have their disk operating system stored in ROM, while others use a short program on the disk itself which is just sufficient to load the rest of the disk operating system automatically, a process known as 'booting DOS'. □ see ²BOOTSTRAP, BUFFER

display *n* any device which presents information in a visual way – **display** *vb* □ see VDU, SEVEN-SEGMENT DISPLAY, LIGHT-EMITTING DIODE, LIQUID CRYSTAL DISPLAY, DOT MATRIX DISPLAY

display file *n* SCREEN MEMORY

distortion *n* anything which corrupts a signal travelling along a communications channel so that less information is received than is sent. Distortion can be caused by faults in the equipment used to send and receive the messages or by outside electrical interference (eg lightning). – **distort** *vb,* **distortional** *adj*

divider *n* **1** – see POTENTIAL DIVIDER **2** a digital counter consisting of a number of flip-flops connected together so as to reduce a frequency by a fixed factor (eg a divide-by-fourteen counter). Dividers are used in digital watches and other devices to obtain pulses at 1 hour, 1 minute, and 1 second intervals from a high frequency crystal oscillator. □ see PRESCALER

DMA – see DIRECT MEMORY ACCESS

documentation *n* **1** a set of instructions (eg a manual) supplied with a computer explaining how to operate the computer **2** a description of how a computer program works

Dolby B noise reduction *n* a method of reducing the level of 'hiss' from audio tapes during quiet passages. The method involves artificially boosting high frequencies during quiet passages in the recording. The converse occurs during replay. A decoding circuit, usu based on an integrated circuit, de-emphasizes the high frequency signals resulting in the required overall flat frequency response, but with the background hiss considerably reduced for low level signals

dongle *n* a device for protecting computer software from illegal copying. A dongle is plugged into a computer and carries a unique coding which is interrogated by the software. If the dongle is not present, or if its coding is different from the serial number embedded in the software, the program cannot be loaded into the computer.

donor *adj* of or being impurity atoms (eg phosphorus atoms), introduced into pure silicon to provide an excess of free electrons – compare ACCEPTOR – **donor** *n* □ see N-TYPE, IMPURITY, DIFFUSION

dopant *n* a substance (eg phosphorus) which is introduced in small quantities into a crystal of silicon so as to produce an n-type or p-type semiconductor □ see VALENCY, N-TYPE, P-TYPE

dope *vb* to produce an n-type or p-type semiconductor by adding an impurity to silicon □ see DOPANT, BIPOLAR TRANSISTOR

DOS – see DISK OPERATING SYSTEM

dot matrix display *n* a method of forming characters by selectively illuminating a matrix of light-emitting diodes or by activating a grid of pixels on a screen □ see DOT MATRIX PRINTER, CHARACTER GENERATOR

dot matrix printer *n* a printer that uses a printhead which forms the shapes of characters as a matrix of small dots. The printhead might be a set of needles which are driven, under computer control, against the paper.

double-density disk *n* a floppy disk which holds twice as much data as a single-density disk. The total amount of data which can be stored on a floppy disk varies from about 90 kilobytes for a single-sided single-density disk to about 1.2 megabytes for a double-sided double density disk. □ see FLOPPY DISK

double-sided disk *n* a floppy disk which can store data on both sides and therefore hold twice as much data as a single-sided disk □ see DOUBLE-DENSITY DISK

downlink *n* – see EARTH STATION

download *vb* to transfer programs sent by the telephone or TV channel to a computer in the home or office – **downloader** *n* □ see MICRONET 800

down time *n* a period of time during which a computer system is not available for use, esp because of a fault in the system which renders it inoperable – compare UP TIME

drain *n* the one of the three terminals of a field-effect transistor which allows current to enter the channel of the device and flow

towards the source terminal □ see SOURCE, FIELD-EFFECT
TRANSISTOR

DRAM [*dynamic random access memory*] a solid-state memory
device based on metal-oxide semiconductor (MOS) technology.
Since the first 1 kilobit DRAM was introduced in the early 1970s,
the number of memory cells (based on individual transistors) on a
memory chip has doubled every year, culminating in the latest 256
kilobit devices. Each one of the 256 thousand memory cells is very
simple and consists of a small capacitor which stores charge and an
n-channel transistor which is switched on to read this charge. Each
memory cell in a DRAM needs to be refreshed from time to time to
boost the small charge on each capacitor which tends to leak away.
Thus a DRAM will not retain its data if the power supply to it is
switched off, ie it is 'dynamic'. □ see RANDOM-ACCESS MEMORY,
NMOS, REFRESH, DYNAMIC, MOSFET

drift *n* **1** a gradual change in the reading of an instrument caused by
temperature changes and/or ageing of components <*station ~ on a
radio tuning scale*> **2** the relatively slow movement of electrons and
holes in a semiconductor caused by temperature or potential
difference – **drift** *vb*

¹drive *n* the motive power of a recording medium □ see DISK DRIVE

²drive *vb* to operate a device by connecting it to control signals or to
a power supply <*to ~ a display*>

driver *n* any device which gives a signal of sufficient power to
operate something (eg a light-emitting diode) □ see BUS DRIVER,
DARLINGTON PAIR

droid *n* a male or genderless android □ see GYNOID, ANDROID

drone *n* a radio-controlled aircraft, missile, submarine, or ship used
for target practice or for remote exploration (eg on the sea floor)

D-type connector *n* a D-shaped plug and socket consisting of two
rows of terminals for connecting printers or other peripheral devices
to computers and other electronic equipment. The plug is normally
attached to the device by a ribbon cable and the socket is built into
the back or side of the computer. D-type connectors contain 5, 25,
or 37 terminals depending on the application. □ see RIBBON CABLE

D-type flip-flop a type of flip-flop (or bistable) which has a 'data'
(D) input so that on every clock pulse the value, 0 or 1, on this input
is output from the flip- flop's output. The JK flip-flop can operate as
a D-type flip-flop. □ see JK FLIP-FLOP

dual-in-line *adj* (abbr **DIL**) of or being a popular integrated circuit
package having terminal pins in two parallel rows, one along each
side of the package. The number of terminal pins varies from 8 to
more than 30 and, in all cases, the separation of adjacent pins along
a side of the package is 2.54 millimetres (1/10 inch). □ see FLAT
PACK, QUAD-IN-LINE

dumb terminal *n* a computer terminal that merely receives data
which the user cannot modify. A VDU displaying flight departures

in an airport lounge is a dumb terminal – compare INTELLIGENT TERMINAL

dump *vb* to transfer a large amount of data from one storage medium to another (eg from RAM to disk) – compare DOWNLOAD – **dump** *n*

duplex *adj or n* FULL DUPLEX

duty cycle *n* the ratio of the time that a rectangular wave signal is 'high' to the total time for a complete cycle of the wave <*the ~ of a square wave is 0.5 or 50 per cent*> – compare MARK-TO-SPACE RATIO

dynamic *adj, of a computer memory* requiring regular and brief 'refresh' signals to retain its contents – compare STATIC □ see DRAM

dynamic dashboard *n* a car dashboard having a computer-controlled readout of speed, fuel level, engine temperature, condition of brakes, etc. A dynamic dashboard makes use of light-emitting diodes and other types of displays to give both analogue and digital information. □ see ELECTRONIC ENGINE MANAGEMENT

dynamic RAM *n* DRAM

E

E the symbol used in the hexadecimal system for the decimal number 14 (eg the hexadecimal number E9 equals the decimal number 14 × 16 + 9 = 133) □ see HEXADECIMAL

EAROM [*e*lectrically-*a*lterable *r*ead-*o*nly *m*emory] – see READ-ONLY MEMORY

earpiece *also* **earphone** *n* a small device plugged into the ear for converting electrical signals into sound waves and used with deaf aids, personal radios, etc. There are two main types of earpiece: the high impedance crystal earpiece in which mechanical vibrations are produced in a quartz or ceramic crystal by applying the electrical signals to two metal terminals attached to the crystal; and the low impedance magnetic earpiece in which mechanical vibrations are produced in an iron diaphragm near to a coil of wire carrying the electrical signals. □ see LOUDSPEAKER, MICROPHONE, TRANSDUCER

¹earth *n* also **ground** a terminal or conducting path in a circuit which is connected to zero potential (ie 0 volts). The connection may be to earth, as in a radio receiver, or it may be to the circuit's chassis or other large area of metal taken to be at zero volts.

²earth *vb* to make a connection to earth or to that part of a circuit which is taken to be at zero volts <*to ~ a radio aerial*>

Earth station *n* a base for relaying radio signals to and from artificial satellites and spaceprobes. An Earth station usu has one or more movable dish-shaped aerials pointing skywards. Signals sent to a satellite are said to be 'uplink' and those coming from the satellite are said to be 'downlink' <*the ~ at Goonhilly Downs in Cornwall*>
□ see COMMUNICATIONS SATELLITE, INTELSAT, DIRECT BROADCAST SATELLITE

ECG – see ELECTROCARDIOGRAPH

echo *n* a reflected signal. Echoes are unwanted in some communications systems since they cause interference (eg 'ghosting' on a VDU). But the echo of radio signals from the Moon is used by amateur radio enthusiasts for long distance communication round the Earth and radar and sonar both use radio echoes to good effect.
– **echo** *vb* □ see GHOSTING, RADAR

ECL – see EMITTER-COUPLED LOGIC

eddy currents *n* circulating currents induced in a conductor by a nearby changing magnetic field. Eddy currents represent wasted electrical energy and the laminations of the core of a transformer are one way to reduce this energy loss. However, eddy currents are put to good use in an induction furnace which uses a strong varying magnetic field to melt metals. □ see LENZ'S LAW, ELECTROMAGNETIC INDUCTION, TRANSFORMER

edge connector *n* a single or double row of contacts in a plastic strip which can be plugged into a printed circuit board to enable circuits to be connected together. The main printed circuit board of some microcomputers has a row of parallel conducting tracks at its edge for an edge connector so that extra memory, joysticks, and other peripherals can be connected. □ see INPUT/OUTPUT PORT, PERIPHERAL

edge triggering *n* a method of making flip-flops in digital counters change state by designing them to respond to the upward rise (positive edge triggering) or downward fall (negative edge triggering) of a clock pulse – compare LEVEL TRIGGERING □ see CLOCK, PULSE

edit *vb* to improve a program by correcting mistakes and inserting new instructions □ see EDITOR

editor *n* a program for creating or changing text stored in a computer file (eg a disk). An editor is also used in a wordprocessor to replace, delete, add, and shift passages of text before printing. □ see WORDPROCESSOR, TEXT EDITOR

educational program *n* software which is intended to teach something useful (eg how to do calculus) □ see ADVENTURE PROGRAM, COMPUTER-AIDED LEARNING

EEM – see ELECTRONIC ENGINE MANAGEMENT

EEROM [*electrically-erasable read-only memory*] – see READ-ONLY MEMORY

effective value *n* ROOT MEAN SQUARE VALUE

efficiency *n* the capacity of a device or system for producing the greatest output for the least input. In electronics, efficiency generally refers to a device's processing ability measured against the electrical power needed to operate it *<a valve has less ~ than a transistor>*. Efficiency is properly defined as the ratio of the power output to the power input and is expressed as a percentage *<the ~ of a loudspeaker is the ratio of the sound power output to the electrical power input>*.

EFT – see ELECTRONIC FUNDS TRANSFER

eight-bit computer *n* a computer, esp a microcomputer, which handles binary data as groups of 8 binary digits □ see BYTE, MICROCOMPUTER, MICROPROCESSOR

electricity *n* (the effects and use of) electrically charged particles (eg electrons and protons) at rest or in motion. An electric current is a flow of such particles. □ see STATIC ELECTRICITY

electrocardiogram *n* the pattern displayed on a VDU or printer by an electrocardiograph □ see ELECTROCARDIOGRAPH

electrocardiograph *n* (abbr **ECG**) a medical instrument that records the electrical signals produced by the heart and picked up by electrodes attached to a person's chest. An electrocardiograph is one of several instruments used in a portable microprocessor-controlled patient monitoring system for measuring a range of body functions including respiration rate, blood pressure, and body temperature. □ see PACEMAKER, BIOCHIP

electrode *n* a metal conductor used to make electrical contact with a circuit <*the* ~s *of a battery*>

electroluminescence *n* the conversion of electricity to light at a relatively low temperature (eg the emission of light by the phosphor on a VDU when struck by an electron beam) □ see THERMIONIC EMISSION, LIGHT-EMITTING DIODE

electrolytic capacitor *n* a capacitor consisting of two aluminium plates separated by a very thin insulating layer of aluminium oxide. An electrolytic capacitor is polarized, which means that it must be connected in a circuit with its positive terminal connected to the more positive voltage. Electrolytic capacitors are used in circuits because they offer a high capacitance in a small volume, but they have the disadvantage that they have a wide tolerance. However, they are used for smoothing varying voltages in power supplies and for other applications where the value of a capacitor is not important. □ see RESERVOIR CAPACITOR, POWER SUPPLY

electromagnet *n* a coil of wire wound on an easily magnetized material (eg iron) which becomes magnetic when a current flows through the coil. The electromagnet is the basis of electromagnetic relays, solenoids, transformers, etc. □ see ELECTROMAGNETIC INDUCTION

electromagnetic induction *n* any means of making (inducing) a current flow in a circuit by changing the strength of a magnetic field linked with that circuit. Transformers, generators, and radio aerials are examples of devices which depend on electromagnetic induction for their operation. □ see LENZ'S LAW, TRANSFORMER, INDUCTANCE

electromagnetic pulse *n* (abbr **EMP**) an intense and sudden burst of electromagnetic energy produced by the explosion of a nuclear weapon in the atmosphere. In addition to the appalling effects of radiation, fire, and blast damage when nuclear weapons are used, the electromagnetic pulse would cause worldwide disruption to radio communications and irreparable damage to many types of electrical equipment. Consequently, military planners see fibre optics cables as one solution to the electrical problems produced by the electromagnetic pulse since laser light carrying information down a fibre optics cable is not susceptible to these effects. □ see OPTICAL COMMUNICATIONS, ELECTROMAGNETIC WAVE

electromagnetic spectrum *n* the entire range of radiations which travel at the speed of light and extend from very short gamma rays to the longest radio waves □ see ELECTROMAGNETIC WAVE, SPECTRUM

electromagnetic wave *n* any wave which travels through a vacuum at the speed of light and consists of oscillating magnetic and electrical fields. Electromagnetic waves include radio, infrared, visible light, ultraviolet, X-rays, and gamma rays. These waves are characterized by their different wavelengths (or frequencies), the means by which they are generated and detected, and the ways in which they interact with matter. □ see RADIO, TELEVISION, OPTOELECTRONICS

electromechanical *adj* depending on a combination of electrical and mechanical functions <*an ~ relay*>

electromotive force *n* (abbr **EMF**) also **electronmoving force** the electrical force generated by a cell or battery which causes a current to flow through a conductor connected between the terminals of the battery. EMF is measured in volts, and is defined as the energy gained by one coulomb of electricity in passing through the battery. □ see POTENTIAL DIFFERENCE

electron *n* a particle which carries a negative charge and which is one of the basic building blocks of all substances. An electric current through a conductor (eg copper) is the movement of a large number of electrons. □ see ELECTRONICS, PROTON, ELECTRICITY, N-TYPE

electron beam *n* a stream of electrons moving in a vacuum under the control of an electric and/or magnetic field □ see CATHODE-RAY TUBE, VDU, SCANNING ELECTRON MICROSCOPE, TELEVISION

electron-beam lithography *n* a method currently under development for obtaining finer circuit detail on the surface of silicon chips than is possible with optical techniques. By using a very narrow electron beam it is possible to define details (eg the width of metal connections) as small as about 0.5 microns. □ see PHOTOLITHOGRAPHY, MICRON, X-RAY LITHOGRAPHY

electron gun *n* a device which accelerates and focusses electrons into a fine beam to produce a spot of light on the screen of a TV, cathode-ray oscilloscope, radar, VDU, or other display device. The electron gun contains electrodes to which high voltages are applied to draw electrons from a heated cathode, and which also control the brightness of the spot of light. □ see CATHODE-RAY TUBE, THERMIONIC EMISSION, RGB GUNS

electronic *adj* of, being, or using devices (eg integrated circuits) to control electricity in a useful way – **electronically** *adv* □ see ELECTRONICS

electronic banking *n* – see ELECTRONIC FUNDS TRANSFER, ELECTRONIC CASHPOINT, TELEBANKING, TELESHOPPING

electronic calculator *n* an electronic device which does arithmetic. Most electronic calculators are based on a single (MOSFET) integrated circuit and the results of calculations are displayed on a seven-segment display (eg a liquid crystal display). Once capable of doing little more than add, subtract, and multiply numbers, most modern electronic calculators have a memory to hold data being used in calculations. Some of them are programmable so that certain lengthy calculations (eg those needed by an engineer) can be carried out automatically under the control of a program built in or entered by the user. Others can be linked to a printer for a hard copy of the results of calculations. The more sophisticated electronic calculators include alarm, clock, and calendar functions, and act as a memo pad.

electronic camera *n* a device using electronic devices and circuits for capturing images on film. In modern electronic cameras, aperture, shutter speed, focussing, and other functions are controlled automatically by decisions taken by what amounts to a built-in computer. Current developments show that it is possible to replace the conventional film in still cameras by a magnetic disk to store the image. Colour prints or video images on a VDU can be obtained from the disk. □ see CHARGE-COUPLED DEVICE, VIDEO CAMERA

electronic cashpoint *n* a facility offered by most banks which allows a person to draw money from his/her account via a cash dispenser installed outside the bank. An electronic cashpoint is connected to the bank's main computer, access to which is gained by putting a plastic card into the dispenser and keying in a personal identity number (PIN). The main computer checks the coding on the card and the PIN and tells the dispenser to provide the money asked for. Sometimes the cashpoint has a printer which provides a printed statement of the money left in the account after the transaction. □ see SMART CARD, CASHLESS SOCIETY, TELESHOPPING, TELEBANKING

electronic component *n* any device (eg a capacitor) which is used in circuits to control the flow of electricity in a useful way

electronic engineering *n* a branch of engineering (eg telecommunications) dealing with the application of electronic devices and systems □ see MICROELECTRONICS

electronic engine management *n* (abbr **EEM**) a microprocessor-based system for providing the most cost-effective motoring for a particular vehicle. A typical EEM system automatically regulates the fuel mixture to suit varying driving demands, cuts off fuel during deceleration, reduces engine idle speed, and controls an electronic ignition system, all of which results in an improvement in fuel consumption and engine efficiency. To provide these control functions a microprocessor receives and processes information about engine temperature, road speed, engine speed, rate of fuel flow, etc, and controls the action of solenoids and stepping motors. □ see TRIP COMPUTER

electronic funds transfer *n* (abbr **EFT**) the taking of money from one account (eg in a bank) and putting it into another account (eg in a building society) by electronic means □ see TELEBANKING

electronic ignition *n* electronic circuits used to time the ignition of the petrol/air mixture in an internal combustion engine. The main feature of the various types of electronic ignition in use in cars and other vehicles is the replacement of the mechanical contact breaker by an electronic sensor. This sensor uses light or magnetism to detect the right moment to send a pulse to the ignition coil, which in turn produces the spark to fire a particular cylinder. The main advantages of electronic ignition over conventional (Kettering) ignition are greater reliability, since there is no contact breaker to erode or wear out, easier control of timing, a good spark at low engine revs, and, so it is claimed, more miles per gallon. □ see ELECTRONIC ENGINE

MANAGEMENT

electronic language translator *n* a device, similar in appearance to a pocket calculator, for translating words and phrases from one language to another. The words to be translated are entered on the keypad and the translated words stored in an internal memory appear on an alphanumeric display. Some types of translator also speak the words displayed.

electronic mail *n* a paperless method for preparing, transmitting, and delivering messages using the storing and processing power of computers. The well-established telex and facsimile services are examples of electronic mail. Newer services are being offered by Viewdata. Among the advantages of electronic mail is the fast delivery of messages, the saving on energy since conventional mail services have to move enormous quantities of paper, and the ease with which messages can be stored, retrieved, and directed to any destination. □ see VIEWDATA, WORDPROCESSOR, ELECTRONIC OFFICE

electronic micrometer *n* a battery-operated instrument with a liquid crystal digital display, used for measuring thickness. The electronic micrometer is an example of how microelectronics can transform the design of a traditional engineering instrument. The digital display makes the instrument easier to use by unskilled people, and it is also more accurate, giving measurements to the nearest millionth of a metre. □ see LIQUID CRYSTAL DISPLAY

electronic money *n* any device (eg a credit card) which is used to buy something without using coins or notes □ see SMART CARD

electronic music *n* the use of electronic devices and circuits to produce agreeable sounds. The process of making music electronically is called music synthesis since sounds are built up by bringing together several electrical signals using devices called synthesizers. In this way, it is possible to imitate the sounds of many different instruments using one device. Some of these keyboards have the facility to read into their memory music which has been encoded onto paper as bar codes. Once in memory, the music can be reproduced or altered by the player and, by linking the device to a home computer, advantage can be taken of the screen and the extra memory available. In fact, most home computers have software packages (eg a ROM cartridge) for making electronic music. For the budding musician, a package might display a musical stave on the screen. Notes can be placed on the stave and played using the computer's own keyboard or by touching the notes on the screen using a light pen or the finger. Professionally produced electronic music uses digital synthesizers which are especially powerful for creating musical effects. By recording a sound on tape as a sequence of binary digits, it is easy to manipulate the data and to produce new sounds, often using the machine's own composition language which is rather like Basic. Although the Musician's Union is trying to ban

the use of programmable music synthesizers, since it fears that
conventional musicians may become redundant, electronic music is
here to stay and it will become increasingly easy to use on home
computers and music synthesizers. □ see ADSR, SPEECH SYNTHESIS,
DIGITAL RECORDING, WAVEFORM GENERATOR

electronic nurse *n* a microcomputer-based system for looking after
a patient's well-being in hospital. One type of electronic nurse
monitors heart-rate, respiration rate, temperature, and blood
pressure and provides a warning should all not be well. A similar
system looks after premature babies who have difficulty breathing
and automatically controls the flow and pressure of the air they
breathe.

electronic office *n* an office which uses electronic equipment to
produce, process, store, and deliver information. An electronic
office makes use of microelectronic devices including
wordprocessors, videotext services, and voice input/output devices
linked to keyboards, printers, VDUs, and other input and output
devices. □ see WORDPROCESSOR, ELECTRONIC MAIL

electronic pen *n* – see LIGHT PEN, DIGITIZER

electronics *n* the study and application of the behaviour and effects
of electrons in transistors, television tubes, and other devices □ see
MICROELECTRONICS

electronic sewing machine *n* a sewing machine based on a
microprocessor that controls the timing and sequencing of stitching
operations. A typical machine provides a large number of functional
and decorative stitches, button hole sequences, and selectable stitch
lengths, each at the touch of a button. The use of the microprocessor
reduces the number of mechanical parts greatly, making the product
more compact, more reliable, and simpler to maintain

electronic sweetening *n* the addition of 'canned' laughter tracks to
a video film of a TV show to add atmosphere and to encourage
viewers' interest in the show

electronic system *n* a set of electronic building blocks (eg an
amplifier) connected together to produce a useful function

electronic toy *n* any electronic device which is entertaining to play
with. More and more electronic toys have microprocessors in them
so that they can be programmed via an on-board keypad. In this way
a free-ranging vehicle can be instructed to negotiate obstacles or a
game can be devised to try and outwit the player by flashing lights or
asking questions in an electronically synthesized voice. Many
electronic toys, esp the microcomputer, are educational and test the
player in response to answers entered on a keyboard or keypad.
Chess and spelling tests are regarded as electronic toys.

electronic typewriter *n* a typewriter which uses microelectronic
devices to provide facilities such as memory for storing text, screen
display to help with corrections, deletions, and overtyping, and an
interface to enable it to be used as a printer for a microcomputer

electron lens *n* a device for focussing a beam of electrons using electric and magnetic fields *<an ~ in an electron microscope>* □ see ELECTRON OPTICS

electronmoving force *n* ELECTROMOTIVE FORCE

electron optics *n* a branch of electronics dealing with the theory and practice of focussing and deflecting a beam of electrons using electric and magnetic fields □ see SCANNING ELECTRON MICROSCOPE

electron volt *n* (symbol **eV**) a unit for measuring the energy of atomic particles, esp the energy which binds together the building blocks (eg electrons and protons) of atoms. The electron volt is defined as the energy gained by an electron moving through a potential difference of 1 volt and is equal to 1.602×10^{-19} joules. □ see ENERGY GAP

electrosensitive printer *n* a printer for producing program listings, data, and graphics characters that uses a printhead which generates sparks to evaporate aluminium from an aluminium-coated paper under the control of a computer

electrostatic *adj* of, being, or producing stationary electrical charges □ see ELECTROSTATIC DISCHARGE, ELECTROSTATIC PRINTER

electrostatic discharge *n* (abbr **ESD**) a potentially damaging flow of static electricity into the terminals of a semiconductor device, esp metal-oxide semiconductors. To guard against damage, metal-oxide semiconductors are transported with their terminal pins embedded in conducting foam or aluminium foil which provides an easy discharge path. Circuit designers avoid wearing clothes made from manmade fibres because static electricity builds up easily on these materials. Some metal-oxide semiconductor devices have diodes built into them to help discharge static electricity harmlessly. □ see COMPLEMENT-ARY METAL-OXIDE SEMICONDUCTOR

electrostatic printer *n* a non-impact printer (eg the Sinclair printer) which burns away a thin metallic coating on the printing paper. An electrostatic printer has one or two single-wire printing heads which generate a spark as they are moved back and forth across the paper. Electrostatic printers are moderately noisy and slow in operation and the paper they use is more expensive than that used for dot matrix and daisywheel printers □ see DOT MATRIX PRINTER, DAISYWHEEL PRINTER

electrostatics *n* a branch of electronics which deals with the effects and uses of electricity at rest (ie static electrical charges). The action of a capacitor is explained by electrostatics □ see CAPACITOR

elegant *adj, of a computer program* stylishly written but not necessarily useful – **elegantly** *adv*, **elegance** *n*

element *n* **1** any component which is part of a larger system *<a transmitter is an ~ of a communications system>* **2** any of more than 100 natural (eg silicon) and manmade (eg einsteinium) substances which are made up of atoms of only one type

embed *vb* to write a computer language so that it forms an integral part of another language □ see LISP

EMF *also* **emf** – see ELECTROMOTIVE FORCE

emitter *n* one of the three terminals of a bipolar transistor. In an npn transistor it is the terminal from which current leaves the transistor. □ see BIPOLAR TRANSISTOR, COLLECTOR, BASE

emitter-coupled logic *n* (abbr **ECL**) a very fast type of logic circuit in integrated circuit form which uses bipolar transistors operating in saturation. The high power dissipation of ECLs, compared with other logic circuits (eg low-power Schottky TTL), makes them unsuitable for high density integrated circuits. □ see COMPLEMENTARY METAL-OXIDE SEMICONDUCTOR LOGIC, TRANSISTOR-TRANSISTOR LOGIC

emitter follower *n* a bipolar-transistor amplifier in which the output signal is taken from across a resistor connected in its emitter lead. The output signal 'follows', ie has the same sign as, the input voltage applied to the base terminal. The emitter follower is useful for driving low impedance loads such as motors and lamps. □ see SOURCE FOLLOWER, COMMON EMITTER AMPLIFIER

EMP – see ELECTROMAGNETIC PULSE

emulator *n* software or hardware which allows one computer to execute the same programs as another (usu more powerful) computer – **emulation** *n*

enable *vb* to activate a microprocessor or other digital electronic device so that it will respond to data fed into it – compare DISABLE

encapsulation *n* the package in which an integrated circuit or other device is protected *<the plastic ~ of an integrated circuit>* – **encapsulate** *vb* □ see PACKAGE, EPOXY

encoder *n* any device which converts information into a form suitable for electronic processing – compare DECODER – **encode** *vb*

energize *vb* to activate a device or a circuit by applying electrical power to it *<to ~ a relay>*

energy gap *n* also **forbidden gap** the energy required by an electron to move between the valency shell and conduction band of an atom. The energy gap for silicon is greater (1.2 electron volts) than for germanium (0.72 electron volts) which gives silicon an advantage over germanium for making transistors and integrated circuits. Silicon devices can operate at higher temperatures than germanium devices, and they are less prone to damage as a result of thermal runaway. □ see CONDUCTION BAND, VALENCY, THERMAL RUNAWAY, P-TYPE, N-TYPE, ELECTRON VOLT

enhancement *n* the improvement of a computer's performance (eg screen resolution) by hardware or software changes □ see UPGRADE

ENIAC [*e*lectronic *n*umerical *i*ntegrator *a*nd *c*alculator] a computer built at the University of Pennsylvania in the 1940s. ENIAC filled a room, was worked by 18 000 valves, needed 200 kW of electricity, weighed 30 tonnes, and cost a million dollars. Today's pocket calculator uses miniature large-scale integrated circuits, weighs a few grams, is battery- powered, and is affordable by almost everyone. □ see COMPUTER GENERATIONS, VALVE

envelope *n* the shape of the amplitude changes of a waveform (eg of a sound wave) from the instant it starts until it has disappeared. Some microcomputers are fitted with sound generation circuits which allow the user to adjust the envelope of a sound to provide specific musical notes and chords. An envelope is usu divided into three main parts: attack, decay, sustain and release. □ see ADSR, ELECTRONIC MUSIC

environment *n* the surroundings in which an electronic device operates *<the ~ of interplanetary space>* – **environmental** *adj*

epitaxial layer *n* a thin layer (a few microns thick) of one type of silicon formed on a layer of another type of silicon. The epitaxial layer is masked, etched, and metallized to produce a transistor or other semiconductor device on a silicon chip □ see EPITAXY, SUBSTRATE, PHOTOLITHOGRAPHY

epitaxy *n* the growth of one type of crystalline silicon on another. Epitaxy is an important process in the manufacture of integrated circuits. A thin layer (a few microns thick) of n-type or p-type silicon is grown on another type of silicon, called the substrate, by heating the substrate in silicon chloride to provide the growth, and in phosphene gas (to give p-type silicon) or in boronethane gas (to give n-type silicon). This layer is then masked and etched to produce semiconductor devices on the silicon layer. – **expitaxial** *adj* □ see SUBSTRATE, PHOTOLITHOGRAPHY, EPITAXIAL LAYER

epoxy *n* a black resinous plastic material used to encapsulate most of the cheaper integrated circuits and transistors, but not used for military and space applications where greater toughness and reliability is expected from the encapsulation

EPROM [*e*rasable *p*rogrammable *r*ead-*o*nly *m*emory] a type of read-only memory (ROM) which once programmed can be reprogrammed by first erasing its contents and then writing in a new program. EPROMs incorporate a window of silica through which ultraviolet light is allowed to pass to the interior of the chip causing the capacitors that store the data to lose their charge. □ see READ-ONLY MEMORY

equalization *n* the process of correcting the frequency response of an amplifier to compensate for deviations in the characteristics of input devices (eg a hi-fi pickup) or output devices (eg loudspeakers) □ see FREQUENCY RESPONSE

equivalence gate *n* EXCLUSIVE NOR GATE

erase *vb* to get rid of information in a storage medium *<to ~ a magnetic tape>* – **erasable** *adj* □ see ERASE HEAD, READ-ONLY MEMORY

erase head *n* that part of a magnetic tape drive unit (eg a cassette recorder) which removes recorded information from the tape □ see READ/WRITE HEAD

ergonomics *n* a study of ways of making machines more pleasant to work with. For example, there are many ways of creating a comfortable workstation: adequate lighting of the keyboard using an adjustable desk lamp, reduction of screen glare from VDUs using coloured filters or polarizing filters, correct height of keyboard, good keyboard layout, and so on. While most microcomputers use the QWERTY layout many computer designers would do away with keyboards entirely. Future computers may well only have a joystick, a mouse, or a touchscreen to make it easier to input data into the computer. □ see MAN-MACHINE INTERFACE, SYSTEMS ANALYST

error message *n* a note or code appearing on a VDU during programming indicating that an error has been made in writing an instruction or in loading a program

ES – see EXPERT SYSTEM

escape *also* **escape key** – see BREAK KEY

ESD – see ELECTROSTATIC DISCHARGE

etching *n* **1** the process of removing silicon oxide from minute areas of the surface of a silicon chip using a solution of hydrofluoric acid, a photomask having been used earlier to define these areas. Dopants are diffused through the gaps left after etching so that regions of p-type and n-type material, which are to become transistors, diodes, etc, are created in the silicon. Etching is an important part of the process of photolithography which produces a miniature circuit on a silicon chip. □ see PHOTOLITHOGRAPHY, PHOTOMASK, TRIKE, ULTRAVIOLET **2** the process of removing a film of copper from the surface of a printed circuit board using an acid solution of ferric chloride. The areas of copper which are to remain on the board as inteconnections between components are marked out using paint, ink, or transfers. A commercial process for making printed circuit boards involves the use of ultraviolet light to expose areas of photoresist which have not been covered by the transfers. The copper under the uncovered areas of photoresist is easy to remove with the acid solution. □ see PRINTED CIRCUIT BOARD, ULTRAVIOLET

Ethernet *trademark* a local area network capable of linking several hundred computers together by means of coaxial cables □ see LOCAL AREA NETWORK

Euroconnector *n* an internationally agreed standard for connecting together video and audio equipment

eV – see ELECTRON VOLT

event *n* an occurrence of something <*the completion of a control program is an ~*>

exclusive NOR gate *n* also **equivalence gate, XNOR gate** a decision-making digital building block which produces an output of binary 1 when both of its two inputs have a value of binary 1 or binary 0. The exclusive NOR gate is used in circuits for comparing two binary digits – compare NOR GATE □ see GATE 1

exclusive OR gate *n* also **XOR gate** a decision-making digital building block which produces an output of binary 1 when one but not both of its two inputs has a value of binary 1, and an output of binary 0 when both its inputs have a value of binary 1 or binary 0. Exclusive OR gates are generally available in integrated circuit packages and form the basis of adding circuits in computers and calculators – compare OR GATE □ see GATE 1

execute *vb* to carry out one or more instructions in a computer program – **execution** *n* □ see EXECUTION TIME, MACHINE CYCLE

execution time *n* the time taken for a computer to work through a set of instructions □ see MACHINE CYCLE

expandable *adj* capable of being extended to fit a range of applications <*an ~ audio system*> – **expand** *vb* □ see PERIPHERAL, MOTHERBOARD, BACKPLANE, EDGE CONNECTOR

expansion bus *n* the set of electrical connections leaving a computer at its expansion port

expansion port *n* a socket or edge connector on a microcomputer for plugging in additional memory, joysticks, modems, printers, etc □ see EDGE CONNECTOR, BACKPLANE

expert system *n* (abbr **ES**) computer software which holds information on some area of human expertise. An expert system can help a doctor to diagnose illnesses, a chemist to understand chemical reactions, or a geologist to search for oil. Until recently, expert systems were available only on big computers, but some microcomputers have the capacity to operate as an expert system. An expert system does not just act as a database to provide answers to questions entered at the keyboard. It is designed to detect a line of enquiry and therefore provide as much useful information as possible in the shortest time. Advances in speech recognition and speech synthesis will soon enable the home computer to behave more like a human expert – perhaps a talking 'home doctor' who will listen sympathetically to the description of an illness and prescribe the appropriate medicine. □ see DATABASE, ARTIFICIAL INTELLIGENCE

Explorer the USA's first successful Earth-orbiting satellite launched on January 31 1958. Electronic instrumentation aboard Explorer measured the intensity of cosmic rays and of the radiation belts round the Earth, the temperature of the upper atmosphere, and the frequency of collisions with micrometeorites. Like Sputnik, launched a year earlier, Explorer had no means of storing data and therefore transmitted information to Earth by radio continuously. □ see SPUTNIK, TELEMETRY, ARTIFICIAL SATELLITE

exponent *n* a number or symbol written at the upper right of another number to show it is to be multiplied by itself a certain number of times. Thus the number 10^6 bytes means 1 million bytes where 6 is the exponent of 10.

external storage *n* storage of programs and data in devices (eg

floppy disks) outside a main computer system
extrinsic *adj, of current flow through a semiconductor* produced by
the addition of minute amounts of selected impurities to extremely
pure silicon or germanium to give an excess of electrons or holes –
compare INTRINSIC □ see SEMICONDUCTOR, P-TYPE, N-TYPE

F

F 1 the symbol for the unit of electrical capacitance, the farad □ see FARAD **2** the symbol used in the hexadecimal system for the decimal number 15 (eg the hexadecimal number FF equals the decimal number $15 \times 16 + 15 = 255$) □ see HEXADECIMAL

facsimile *n* (abbr **FAX**) the sending by telephone of pages of printed material, including drawings and pictures, using a type of modem connected to the telephone line. The page to be transmitted is scanned in a series of lines so that the light and dark areas are converted into varying analogue or digital signals. The receiving station decodes these signals to produce a copy of the original page. □ see MODEM, ELECTRONIC OFFICE, TELETYPEWRITER

fading *n* the variation in the strength of a radio signal caused by interference between two parts of a signal reaching the receiver by different paths *<~ on a car radio>* – **fade** *vb* □ see ATTENUATION

failsafe *adj* ensuring no loss of function or danger when failure occurs *<battery backup on computer memory is a ~ method of ensuring that stored programs are not lost>*

fall time *n* the time for a signal waveform to fall from 90% to 10% of its maximum voltage – compare RISE TIME

false *n* the state of a digital logic signal represented by binary 0 – compare TRUE □ see LOGIC LEVELS

family *n* – see LOGIC FAMILY

FAMOS [*f*loating-gate *a*valanche-injection *m*etal-*o*xide *s*emiconductor] a type of technology used for making programmable read-only memories and based on storage cells which are similar to field-effect transistors. When a high voltage is applied across the drain and source terminals of a cell, breakdown occurs in the channel and a surge of electrons flows. A number of these electrons become trapped in the gate region forming a charge which can be read later. Ultraviolet light is used to make the charge leak away so that the device can be reprogrammed. □ see METAL-OXIDE SEMICONDUCTOR FIELD-EFFECT TRANSISTOR, READ-ONLY MEMORY

fanfold paper *n* also **concertina fold paper, accordion fold paper** a continuous length of printer paper divided into pages by perforations so that the paper folds first in one direction then in the other to form a pile – compare ROLL PAPER

fan-in *n* the number of logic circuits connected to the input of another logic device *<a ~ of six>* – compare FAN-OUT – **fan-in** *vb*

¹fan-out *vb* to drive a number of devices from a single signal source *<to ~ a logic signal to a number of gates>* – compare FAN- IN

²fan-out *n* the maximum number of logic gate inputs which can be driven by the output signal of a similar logic gate (eg the fan-out of a TTL logic gate is 10)

farad *n* (symbol **F**) the unit of electrical capacitance. If a potential difference of 1 volt across the terminals of a capacitor causes it to store an electrical charge of 1 coulomb, the capacitance of the capacitor is 1 farad. □ see CAPACITOR, COULOMB, CAPACITANCE

fault *n* an unforeseen error arising in equipment or programs – **faulty** *adj* □ see BUG

FAX – see FACSIMILE

FDM – see FREQUENCY DIVISION MULTIPLEXING

feasibility study *n* a look at the possible solutions to a problem (eg how to design 'intelligent' computer systems)

feed *vb* **1** to put information into a device <*to ~ data into a computer*> **2** to pass paper through a printer so that another line can be printed – **feed** *n*

feedback *n* the sending back to the input of part of the output signal of an amplifier so as to make it behave in a useful way □ see NEGATIVE FEEDBACK, POSITIVE FEEDBACK

feedback factor *n* the proportion of the output signal which is fed back to the input of an amplifier □ see FEEDBACK

feedback loop *n* the path a signal takes in passing from the output to the input of a system <*the ~ of an electronic thermostat*> □ see FEEDBACK, FEEDBACK FACTOR

ferrite *n* powdered iron oxide commonly used, in the form of a hard rod, to increase the sensitivity of a radio aerial coil wound on it and as the magnetizable part of magnetic tapes and disks – **ferrite** *adj* □ see INDUCTOR, CORE MEMORY, MAGNETIC TAPE

ferrite rod aerial *n* a type of aerial used to make radio receivers more sensitive. A ferrite rod aerial is made by winding a coil of wire on a ferrite rod; it provides maximum sensitivity when the rod is lined up end on to the transmitting station – compare DIPOLE AERIAL □ see FERRITE

FET – see FIELD-EFFECT TRANSISTOR

fetch *vb* to retrieve a piece of data from a storage register before processing it □ see FETCH CYCLE

fetch cycle *n* a short period of time in the machine cycle of a microprocessor during which data is transferred from a memory location to a register before it is acted on □ see MACHINE CYCLE

fibre optics *n* the use of hair-thin transparent glass fibres to transmit light by repeated internal reflections in the fibre □ see OPTICAL FIBRE, OPTICAL COMMUNICATIONS

fibre-optics communications *n* OPTICAL COMMUNICATIONS

fibre optics connector *n* a device which enables laser light to be passed from one optical fibre to another without reducing the strength of the signal carried by the laser light □ see OPTICAL FIBRE

field *n* **1** a group of computer instructions dedicated to a particular purpose (eg reading data) **2** the property of the space surrounding charged and/or magnetized bodies which causes one body to act on another <*the electrical ~ of an electron*> **3** a set of scanning lines which builds up a complete TV picture line-by-line

field-effect transistor *n* (abbr **FET**) a three terminal device made
from n-type and p-type semiconductors which is used to switch and
amplify electrical signals. Field-effect transistors are used singly or in
many thousands as the basic building blocks of integrated circuits on
a silicon chip. There are two types of FET: the n-channel FET and
the p-channel FET. The channel is the conducting path between the
drain and source terminals through which current flows under the
control of a potential difference applied across the gate and source
terminals. The gate terminal of a junction-gate FET (JUGFET) is
made from a reverse-biased pn junction which is embedded in the
channel, while that of a metal-oxide semiconductor FET (MOSFET)
is insulated from the channel by a very thin layer of silicon oxide. It
is the MOSFET construction (sometimes known as the insulated-
gate field-effect transistor or IGFET) which is favoured by the
makers of integrated circuits, since it is possible to construct many
thousands of these transistors on a very small area of silicon with the
additional advantage, compared with bipolar transistors, that the
circuit consumes less power – compare BIPOLAR TRANSISTOR □ see
METAL-OXIDE SEMICONDUCTOR FIELD-EFFECT TRANSISTOR
field frequency *n* – see PICTURE FREQUENCY
FIFO [*first in first out*] a method of storing and retrieving
information in a memory such that the first piece of data stored is
the first to be retrieved, ie the data which has been waiting the
longest is the first to be dealt with – compare LIFO □ see QUEUE
fifth generation computer *n* a supercomputer which can acquire
knowledge and use it intelligently. The fifth generation computer
was first proposed as a 10-year project by Japan in 1981 and
subsequently this challenge has been taken up by other countries. Its
success however, depends on finding solutions to five problems: first,
how to build a computer on a chip which is considerably more
complex than today's very large scale integration devices; second,
how to devise a language able to handle over 100 million instructions
per second so that it can make intelligent use of information fed to
it; third, how to provide the computer with an enormous and
complex knowledge base which it can draw upon for its decision-
making tasks; fourth, how to design a computer architecture which
performs many tasks at once (parallel processing) so that it acts in a
way more closely resembling the human brain in making inferences
from the information it receives; fifth, how to improve existing
software such as the language Prolog, so that the computer can
recognize human speech. People will be able to hold a conversation
with a fifth generation computer and we shall be able to consult it as
an expert in any chosen field of knowledge. It will make possible
intelligent robots which can see and hear spoken commands, and
which can be involved in decision-making in government, industry,
and the home. □ see COMPUTER GENERATIONS, ARTIFICIAL
INTELLIGENCE, TRANSPUTER, KNOWLEDGE ENGINEERING, PARALLEL

PROCESSING, VERY LARGE SCALE INTEGRATION, EXPERT SYSTEM, PROLOG

file *n* a collection of programs or data stored under one name on a disk or cassette <*a data ~ stores numerical data while a text ~ stores words and sentences*>

file handling *n* the process of dealing with computer files, including sorting through them and deleting unwanted material

file name *n* the set of characters which identifies a computer file

filter *n* **1** a device for controlling the range of frequencies which passes through a circuit (eg a low-pass filter in a hi-fi amplifier reduces hiss produced by the surface of a worn record) □ see HIGH-PASS FILTER, LOW-PASS FILTER, BAND PASS FILTER **2** a film of transparent material for controlling the type of light passing through it (eg an ultraviolet filter in front of a liquid crystal display prolongs the life of the display by protecting it from ultraviolet light) □ see LIQUID CRYSTAL DISPLAY, POLARIZING FILTER

FINDIP package *n* a type of encapsulation used for medium power dual-in-line integrated circuits which has a metal fin jutting out of each side for soldering to the copper of a printed circuit board where it gets rid of excess heat generated in the package

fire *vb* **1** to press a button (on a joystick) to simulate the shooting of something in a computer game <*to ~ a laser*> **2** to activate a device in a circuit <*to ~ a thyristor*>

fire button *n* also **hit key** a switch, usu attached to a joystick or track ball roller connected to a microcomputer, which when pressed sends a 'fire' signal □ see FIRE

firmware *n* computer instructions stored on a chip (eg a read-only memory) as opposed to software which is stored on magnetic tape or disk □ see SOFTWARE, HARDWARE

first generation computer *n* – see COMPUTER GENERATIONS

first in first out – see FIFO

flag *n* a binary signal (ie a 1 or a 0) which a microprocessor uses to show that a particular operation has been completed. The flags are held in an 8-bit register and each of the bits is controlled separately by the arithmetic and logic unit. If one of these bits is a 1, the flag is said to be 'set' showing that an operation has happened. If the bit is 0, the operation has not happened. □ see CARRY FLAG

flag register *n* also **status register** a special register in a microprocessor which contains a set of flag bits used to record whether a particular condition has occurred (eg whether there is a carry bit after an addition of two binary numbers) □ see REGISTER, FLAG, CARRY FLAG

flat pack *n* an integrated circuit package which has leads sticking out of its sides 'horizontally' (ie in the same plane as the package) rather than 'vertically'. This design allows the leads to be soldered to a printed circuit board more easily than with the dual-in-line package, and the low profile means that high density circuits can be designed.

□ see DUAL-IN-LINE, PACKAGE COUNT

flat screen display *n* – see FLAT SCREEN TV

flat-screen TV *n* (abbr **FSTV**) any TV, esp a pocket-size one, using a TV tube which is much flatter than the usual tube. There are two main methods for making a flat-screen tube. One way is to use the conventional cathode-ray tube but to put the electron gun at the side of the screen and to use electrodes to deflect the beam through 90 degrees producing a picture which is viewed from the same side as the light-emitting phosphor. The second method is not to use a cathode-ray tube but a liquid-crystal display, a method which holds the promise of low-power large screen displays as thin as a picture. □ see CATHODE-RAY TUBE, LIQUID CRYSTAL DISPLAY, PHOSPHOR, WRISTWATCH TV, PORTABLE MICRO

flight recorder *n* a data recorder carried aboard an aircraft to record selected functions and flight patterns of the aircraft for use in case of an accident. First designed in the 1960s, most flight recorders scratch four measurements on a revolving roll of alloy foil. The measurements are: height, heading, speed, and vertical acceleration of the plane up to the time of the accident. Since the paper moves at only 150mm per hour the resolution of the measurements is poor. However, increasing use is being made of advanced digital flight recorders which record up to 64 different readings on magnetic tape. These measurements include important additional facts such as the aircraft's pitch and roll attitudes, the position and motion of the control surfaces (eg rudder, ailerons, and flaps), engine performance, and longitudinal acceleration.

flight simulation *n* a popular game available for most microcomputers which involves the player in piloting an imaginary aircraft. Instruments displaying wind speed, air speed, altitude, fuel level, etc and radar maps and radio beacons are represented on the VDU to help in landing, take-off, and navigation across country.

flip-flop *n* also **bistable, toggle** an electronic building block which has two outputs acting as stores for binary data fed to its input. The name of this single memory cell comes from the fact that the two outputs always have different logic levels (binary 1 and binary 0) and, on receiving appropriate input signals, each output changes from one level to the other. Flip-flops, using the transistor as the basic building block, are used in all types of digital counting circuits (eg in watches), and are the basic memory elements in static random-access memory devices in computer systems. □ see JK FLIP-FLOP, BINARY COUNTER, RANDOM-ACCESS MEMORY

float *vb* to be at an unknown voltage with respect to supply voltage <*the inputs of the chip were* ~*ing*> □ see FLOATING INPUT

floating head *n* the read/write head in a Winchester disk drive which is supported just above the disk on a cushion of air generated by the rapid rotation of the disk □ see WINCHESTER DISK

floating input *n* an input to an integrated circuit which is left unconnected. A floating input may acquire an unknown voltage from either internal or external sources of signal and cause erratic behaviour of the IC. Thus it is recommended that all unused inputs are not left to 'float' but should be connected to either of the power supply connections.

floppy disk *n* also **diskette** a flexible disk made of plastic and coated with a magnetic film on which is stored computer data in digital form. The usual diameter of a floppy disk is 5¼ inches (133 millimetres), and it is enclosed in a protective square envelope pierced with holes for the drive shaft and read/write head. Floppy disks are less reliable but much cheaper than hard disks – compare HARD DISK, WINCHESTER DISK

flow *n* also **stream** 1 a sequence of data <*a ~ of information from a weather satellite*> 2 a succession of events <*a ~ of instructions carried out by computer*> 3 a movement of electrical charge <*current ~*> – **flow** *vb*

flowchart *also* **flow diagram** *n* a diagram which uses symbols connected by arrows to show the sequence of actions necessary to solve a particular problem. For example, a diamond-shaped symbol is used in a flowchart when a choice between two routes is to be made that depends on a yes or no answer to a question. □ see ALGORITHM, SYMBOL

flow diagram *n* FLOWCHART

fluorescent vacuum display *n* a display in which each segment of a number consists of a heated filament which produces electrons that strike a fluorescent screen in an evacuated glass tube to produce a blue-green light – compare GAS DISCHARGE DISPLAY

flutter *n* – see WOW AND FLUTTER

flyback *n* the return of an electron beam in a cathode-ray tube to the starting point after it has completed a trace □ see SCANNING BEAM, CATHODE-RAY TUBE, TELEVISION RECEIVER

FM – see FREQUENCY MODULATION

footprint *n* 1 an area on the Earth's surface covered by the transmissions of a communications satellite □ see COMMUNICATIONS SATELLITE, DIRECT BROADCAST SATELLITE 2 the space taken up by a desk-top computer. Manufacturers of office equipment often make a sales point that their computers and word processors require less office space than those of their competitors. □ see ELECTRONIC OFFICE

forbidden gap *n* ENERGY GAP

foreground colour *n* the colour that text and graphics are printed in on a VDU – compare BACKGROUND COLOUR

foreground processing *n* a processing task a computer turns to automatically in preference to jobs of lesser importance – compare BACKGROUND PROCESSING

format *n* the layout of data on a VDU, a printout, or a storage medium <*the disk ~*> – **format** *vb* □ see FORMATTING, TABULATION

format effector *n* a special symbol (eg BS to mean 'back space') used in a data communications system to inform a printer or other peripheral to arrange data in a particular way

formatting *n* the organization of the memory space on a magnetic disk into tracks and sectors *<~ a disk>*. A minifloppy disk is usu divided into 40 tracks, or 80 tracks if it is double-tracked, and 10 to 16 sectors. A computer using the disks keeps a record of the tracks and sectors and uses it to find data and programs on the disk. – **format** *vb* □ see FLOPPY DISK, TRACK, SECTOR, HARD-SECTORED DISK, SOFT- SECTORED DISK

Forth *n* a programming language developed in the 1960s largely for process control applications. By comparison with Basic, Forth is difficult to learn but programs in Forth run much faster.

Fortran *n* [*for*mula *tran*slator] a high-level computer language developed in the 1950s mainly for mathematical and scientific work. Fortran does not handle words very well and there are a number of variations of the language (eg Fortran 77 is a recent version). □ see HIGH-LEVEL LANGUAGE

forward bias *n* a voltage applied across a pn junction to encourage the flow of electrons across the junction. Forward bias is essential to the operation of the rectifier diode and the bipolar transistor – compare REVERSE BIAS – **forward biased** *adj* □ see PN JUNCTION, BIPOLAR TRANSISTOR

frame *n* a screenful of information on a VDU

FRED [*f*rantically *r*apid *e*lectronic *d*evice] an early example (it is said) of a name chosen as an acronym □ see BUZZWORD

free-standing *adj* also **stand-alone 1** *of hardware* not requiring physical or electrical support from other equipment **2** *of software* capable of being used independently of other software *<a ~ program>*

frequency *n* the number of times per second that a periodic event (eg the reversal of the mains current) repeats itself. The unit of frequency is the hertz. The mains current has a frequency of 50 hertz (50Hz) and sound waves to which the human ear is sensitive have frequencies in the range 20 hertz to 20 kilohertz. □ see CLOCK, ELECTROMAGNETIC WAVE, AUDIO FREQUENCY

frequency division multiplexing *n* (abbr **FDM**) a method of transmitting a number of different speech signals down a single telephone line or other communications channel. Each message modulates a separate carrier frequency separated by about 4 kilohertz. At the receiving end, filters separate each carrier frequency and the message is recovered by demodulation – compare TIME DIVISION MULTIPLEXING

frequency modulation *n* (abbr **FM**) a method of sending information by controlling the frequency of a wave in response to changes in the amplitude of the signal representing the information. For high-quality radio broadcasts, frequency modulation is preferred to amplitude modulation since the transmissions are less affected by interference generated by lightning and other electrical discharges – compare AMPLITUDE MODULATION, PHASE MODULATION

frequency response *n* the ability of a circuit or device to carry or amplify different frequencies. The frequency response of a hi-fi amplifier is a measure of how well it amplifies both high and low audio frequencies. □ see BANDWIDTH

frequency shift keying *n* (abbr **FSK**) a method of transmitting binary information by radio where one carrier frequency is used for 0 and a different frequency for 1

friction feed *n* a method of moving paper through a printer by gripping it between the platen and a roller – compare TRACTOR FEED

front end *n* **1** that part of a computer system operated by the user *<a keyboard is the ~ of a computer>* **2** that part of a communications system that receives incoming signals *<the ~ of a radio is its tuner>* – compare BACK END – **front end** *adj*

front panel *n* a set of lights and control switches on the front of electronic equipment to show how the equipment is working and to enable the user to operate it □ see CONSOLE, FRONT END

FSK – see FREQUENCY SHIFT KEYING

FSTV – see FLAT SCREEN TV

full-adder *n* a digital circuit used in calculators and computers which adds together two two-bit binary numbers and provides a two-bit sum and a carry bit – compare HALF-ADDER

full duplex *also* **duplex** *adj or n* (of or being) a communications system which allows information to flow in opposite directions simultaneously – compare HALF DUPLEX

full travel *adj* of or being a computer keyboard having keys which can be pressed right down – compare MEMBRANE KEYBOARD

full-wave rectifier *n* a semiconductor device using two or four diodes which produces direct current from alternating current by reversing the flow of current in one half cycle of every full cycle of the alternating current – compare HALF-WAVE RECTIFIER □ see POWER SUPPLY, RECTIFIER

function *n* **1** a set of operations in a computer program which produces a result **2** a mathematical relationship between two variable quantities *<the ~ F = 32 + 9 × C/5>* **3** an operation performed by a device *<timing ~>*

function generator *n* WAVEFORM GENERATOR

function key *n* a key on a keyboard which produces an effect devised by the programmer as a result of a short set of instructions

fundamental *n* of or being a tone having the component of largest amplitude and usu the lowest frequency – compare HARMONIC – **fundamental** *adj* □ see ELECTRONIC MUSIC

fuse *n* a device for protecting equipment from excessive current passing through it. A fuse acts as a 'weak link' in the circuit and usu consists of a length of wire stretched or coiled between two terminals. These terminals are connected in series with the circuit to be protected. If the current through the wire exceeds its rated capacity, the wire melts and breaks the current flow. – **fuse** *vb*

fuse PROM *n* a type of programmable read-only memory (PROM) which can be programmed by means of a PROM burner □ see PROM BURNER, READ-ONLY MEMORY

fusible link memory *n* a read-only memory which can be irreversibly programmed by breaking microscopic metallic links with a burst of high current □ see READ-ONLY MEMORY

G

G the symbol for the prefix giga meaning one thousand million
□ see GIGAHERTZ

Gabor, Dennis..Inventor of holography (1949); a photographic
technique for recording and showing three-dimensional images
□ see HOLOGRAM, THREE-D TELEVISION, OPTICAL MEMORY

gadolinium *n* a metallic magnetic element of atomic number 64
which is used for making magnetic bubble memories □ see
MAGNETIC BUBBLE MEMORY

gain *n* the increase in the power, voltage, or current of a signal as it
passes through an amplifier. An amplifier which has a gain of 1000
provides an output signal 1000 times bigger than the input signal.
□ see AMPLIFIER, DECIBEL, CLOSED-LOOP GAIN, OPEN-LOOP GAIN

galactic *adj, of a resource* large and available to widely scattered
users *<a ~ expert system>* – compare GLOBAL

Galileo the name of an interplanetary probe due for launch from
Earth in 1986 and destined to reach the planet Jupiter in 1988. There
it will be placed in orbit by remote control from Earth to study the
atmospheric and surface conditions of this, the largest planet in the
solar system. While in orbit, Galileo will release a probe to take
measurements as it falls slowly through Jupiter's atmosphere on a
parachute. The observations made by instruments on-board Galileo
and relayed from the atmospheric probe will be radioed back to
Earth so that astronomers can learn more about the origins of the
solar system, and about the violent electrical behaviour of Jupiter.
□ see VOYAGER

gallium arsenide *n* a crystalline material which, like silicon, is used
to make diodes, transistors, and integrated circuits. Its main claim to
fame is that semiconductor devices made from it are able to conduct
electricity five to ten times faster than those made from silicon. This
property makes it of interest to manufacturers of computer memory
devices since there is an ever-increasing need for computers to
process information faster. Data held in a computer memory device
made from gallium arsenide can be stored and retrieved more
rapidly than in any other type of computer memory. However, there
are a few drawbacks to the use of gallium arsenide, one of which is
that gallium and arsenic, the two elements from which it is made, are
in short supply, mainly being found as impurities in aluminium and
copper ores respectively. On the other hand, silicon is plentiful being
found in silicates such as sand. The present cost of gallium arsenide
is about thirty times that of silicon. Furthermore, it is not so easy to
manufacture integrated circuits from it as it is from silicon, since it
does not form a protective layer of oxide to resist the diffusion of a
dopant during the process of photolithography. Thus it is unlikely
that there will be a rapid rise in the use of gallium arsenide-based

semiconductors in the foreseeable future except for specialist applications where cost is not a major consideration. At present, therefore, it is being used in discrete devices for operation at high frequency where silicon devices are not fast enough, and for memory devices in military equipment. □ see MAGNETIC BUBBLE MEMORY, LASER DIODE, PHOTOLITHOGRAPHY

gamma counter *n* a microprocessor-based medical instrument for measuring the concentration of substances in a medical sample by detecting the emission of gamma rays from it after it has been made artificially radioactive. A gamma counter has a keyboard, VDU, and printer so that measurements can be recorded in graphical or printed form as required.

Gamma-Ray Observatory (abbr **GRO**) an Earth-orbiting observatory designed to investigate gamma-rays from stars and galaxies and due for launch by the space shuttle in 1988. The shuttle will also retrieve and service the observatory from time-to-time. The observatory will carry four different electronic instruments for measuring the intensity and wavelength of gamma rays from celestial sources. These observations will help astronomers seek out black holes at the centre of violently active galaxies called quasars.

gamma rays *n* electromagnetic radiation of considerably shorter wavelength than light, that is generated mostly by radioactive substances. Gamma rays cause damage to body tissues in high concentration and they are very penetrating so that it is difficult to protect oneself from their effects. □ see GAMMA COUNTER

gap *n* the small separation between a read/write head and a magnetic disk or tape □ see WINCHESTER DISK

garbage *n* incomprehensible data either put into a computer by an incompetent programmer or produced by a computer as a result of a hardware fault or operator error □ see GIGO

garbage collector *n* a program which automatically deletes any discarded data (eg strings of the form 'LET A$=A$+...') from a computer's random access memory to prevent the memory from getting full

garbage in garbage out *n* – see GIGO

gas discharge display *n* a display in which each segment of a number consists of a tube containing neon gas which produces a strong orange light when a voltage of about 150 volts is applied across terminals at the ends of the tube – compare FLUORESCENT VACUUM DISPLAY

gaseous diffusion *n* also **planar diffusion** the process of heating a silicon chip in an oven containing a selected impurity in gaseous form so as to produce precisely defined regions of n-type and p-type silicon in the chip. These regions usu become diodes and transistors of an integrated circuit. The impurity (eg phosphene which gives a p-type semiconductor) diffuses into the silicon through minute windows in a layer of silicon dioxide on the surface of the chip. The windows are marked out using a mask and the processes of photolithography and etching. □ see PHOTOLITHOGRAPHY, P-TYPE, N-TYPE

gas sensor *n* a device used in an electronic alarm system for detecting the presence of harmful or explosive gases such as butane, methane, liquefied petroleum gas, and natural gas □ see SENSOR, TRANSDUCER

¹gate *vb* to control the flow of data through a circuit by opening and closing an electronic switch <*to ~ an electronic counter*>

²gate *n* **1** *also* **logic gate** an electronic switch (eg a NAND gate) which acts as a decision-maker in digital systems by giving a binary output of either 1 or 0 depending on the combinations of binary input **2** one of the three terminals of a field-effect transistor □ see FIELD-EFFECT TRANSISTOR

gate oxide *n* a very thin layer of silicon oxide which is grown on the channel of a metal-oxide semiconductor field-effect transistor during its formation on a silicon chip □ see PHOTOLITHOGRAPHY

gateway *n* an interface between information networks. One type of gateway is the facility which enables users to connect directly to the computer of an information provider in a Viewdata system □ see NETWORK, VIEWDATA, INFORMATION PROVIDER

Geiger counter *n* an instrument for measuring the strength of radiation (eg gamma rays) from radioactive materials using a Geiger tube □ see GEIGER TUBE

Geiger tube *n* a device for detecting radiations (esp gamma rays) from radioactive materials. A Geiger tube is a cylindrical tube containing two electrodes, an anode and a cathode, between which is maintained a voltage of a few hundred volts. The tube contains a small quantity of gas at low pressure. Any radiation entering the tube causes the gas to ionize and brief small pulses of current are produced which are counted by special electronic circuits to give a reading of the strength of the radiation on a meter.

generator *n* **1** a program for producing other programs of a particular type <*a compiler ~*> **2** an electronic circuit (eg a waveform generator) which is designed to produce electrical signals of predetermined shape and frequency □ see WAVEFORM GENERATOR, RAMP GENERATOR

geostationary *adj* also **geosynchronous** of or being an artificial satellite remaining in one part of the sky. To be effective, most communications satellites are placed in a geostationary orbit at an altitude of 35 900km above the Earth's equator. At this altitude they revolve round the Earth at the same rate as the Earth spins on its axis and so they are always accessible from Earth stations. □ see COMMUNICATIONS SATELLITE, DIRECT BROADCAST SATELLITE, EARTH STATION

germanium *n* a nonmetallic element of atomic number 32; the first material to form the basis of transistors and diodes. Like silicon, germanium forms crystals in which neighbouring atoms link together by sharing electrons, a process called covalent bonding. The pure germanium crystal can have its electrical properties modified usefully to produce p-type and n-type semiconductor. However, a germanium atom does not hold on as tightly as a silicon atom to the electrons swirling in orbit round its nucleus. This makes semiconductor devices based on germanium more likely to suffer damage caused by self-heating (or thermal runaway). This is one of the reasons why silicon has largely displaced germanium for semiconductor devices – compare SILICON □ see P-TYPE, N-TYPE, PN JUNCTION, BIPOLAR TRANSISTOR, VALENCY

ghosting *n* **1** the appearance of a faint second image of a printed character caused by bounce of the paper or print head **2** the appearance of a faint second image on a TV caused by signals reaching the receiver by different paths **3** the appearance of a faint second image on a VDU or TV caused by too much contrast or brightness

gigabit *n* a quantity of binary data equal to 1 thousand million (10^9) bits. A compact disk can store about 10 gigabits of music information

gigabyte *n* (symbol **GB**) a quantity of computer data equal to 1 thousand million (10^9) bytes □ see FIFTH GENERATION COMPUTER

gigahertz *n* (symbol **GHz**) a frequency equal to 1 thousand million (10^9) hertz which for radio waves is known as ultra high frequency <*radar waves are in the* ~ *range*> □ see UHF, X-BAND

gigo [*g*arbage *i*n *g*arbage *o*ut] a blunt reminder that even the best of present-day computer systems will not correct the errors of incompetent programmers □ see GARBAGE

glare *n* – see SCREEN GLARE

glitch *also* **glytch** *n* a brief but often recurring distortion to the shape of a signal as seen on a cathode-ray tube or other recording device □ see BUG

global *adj* available to several different programs or users <*a* ~ *variable*> – compare GALACTIC

glytch *n* GLITCH

gold *n* a chemically stable element of atomic number 79 which is a good conductor of electricity and can be drawn into a fine wire. Hair-thin gold wire is often used for connecting a completed silicon chip to the terminals of the package in which it is housed. Reclaiming this gold is one of the reasons why some enterprising people are melting down old computers. □ see ALUMINIUM

golfball *n* a small spherical printhead embossed with characters which is tilted and rotated and brought into a position where it can be struck against the paper to print the selected character <*a* ~ *typewriter*> – compare DAISYWHEEL PRINTER

graph *n* a drawing showing the relationship between two numerical quantities (eg volts and amperes for a conductor). A commonly used graph is a line or a set of bars between two axes at right angles to each other. □ see GRAPH PLOTTER, CHARACTERISTIC, HISTOGRAM

graphics *n* pictorial information that is displayed on a VDU and that can usu be manipulated using a light-pen or keyboard *<computer ~>* □ see GRAPHICS TERMINAL, COMPUTER-AIDED DESIGN

graphics characters *n* small shapes occupying the same space as a letter or number on a VDU and used to improve games programs, graphs, etc. Most microcomputers have predefined graphics characters marked on some keys and these are known as its graphics set. It is usu possible to make up graphics characters to one's own design by controlling individual pixels within a graphics square on the screen. □ see PIXEL

graphics mode *n* a way of operating a computer so that it displays graphics to the best effect. Some microcomputers have more than one graphics mode which offer different resolutions. □ see RESOLUTION

graphics set *n* the group of graphics symbols provided by the manufacturer of a microcomputer and usu marked on a group of keys □ see GRAPHICS CHARACTERS

graphics tablet *n* DIGITIZER 1

graphics terminal *n* a terminal with a screen adapted for displaying high resolution graphics □ see COMPUTER-AIDED DESIGN, LIGHT PEN, MOUSE

graph plotter *n* a device connected to a computer to draw graphs, bar charts, and similar graphics using one or more pens for writing and drawing on paper □ see DIGITIZER 1

Gray code *n* a binary code, used to reduce errors in the measurement of linear or angular movement, in which only one bit changes in moving from one value to the next. Thus the Gray code from zero is: (0000), (0001), (0011), (0010), (0110), and so on. In the normal binary code, all bits change between binary 7 (0111) and binary 8 (1000), for example, which makes a measurement error more likely. Angular movement is often measured by representing the Gray code as a set of concentric rings on a disk attached to the rotating device (eg a wind direction indicator). Each ring is divided into a dark and light part. Light-sensitive cells sense where the light part is on each ring. The output of these cells is passed into a computer or digital counter to enable the angular position of the ring to be measured. □ see BINARY CODE

GRO – see GAMMA-RAY OBSERVATORY

grommet *n* a ring of rubbery plastic which is fitted into a hole to protect a cable, esp a mains cable, passing through it from wear. □ see RELIEF BUSH

ground *n* [1]EARTH

ground wave *n* also **surface wave** a radio wave from a transmitter which reaches a radio receiver by following the curvature of the Earth. The range of the ground wave reduces with increasing frequency of the radio wave, and is less over sand than over water. Thus long waves can travel about 1500 kilometres while VHF waves travel just a few kilometres.

grow *vb* to produce a pure crystal of silicon, gallium arsenide, or other semiconductor from a molten quantity of the same material – **growth** *n* □ see BOULE

gulp *n* a small group of bytes □ see BYTE, NIBBLE

gun *n* – see ELECTRON GUN, RGB GUNS

Gunn diode *n* a special type of semiconductor diode made from n-type gallium arsenide and used for generating microwaves. Gunn diodes are used in communications systems, microwave intruder alarms, and police radar traps. □ see MICROWAVES, GALLIUM ARSENIDE, RADAR

gynoid *n* a female android □ see DROID, ANDROID

H

H the symbol for the unit of electrical inductance, the henry □ see HENRY

hack *also* **hacker** *n* someone with little experience or knowledge of computing who spends a lot of time doing insignificant things with computers – **hack** *vb*

half-adder *n* a digital circuit used in calculators and computers which adds together two one-bit binary numbers and provides a one-bit sum and a one-bit carry – compare FULL-ADDER

half duplex *adj or n* (of or being) a communications system that allows information to flow in either direction alternatively but not simultaneously – compare FULL DUPLEX

half wave rectifier *n* a semiconductor diode, or a circuit based on a diode, which produces direct current from alternating current. The diode or circuit allows only one half of a full wave of the alternating current to pass through it – compare FULL WAVE RECTIFIER □ see ALTERNATING CURRENT, RECTIFIER

Hall-effect switch *n* a miniature on/off switch based on semiconductor materials and operated by a magnet. The Hall-effect switch is fast-acting and requires no physical contact to operate it. This type of switch is useful for measuring speed of rotation (eg of gear wheels) and for counting objects moving along a production line, for example. In the future, the Hall-effect switch is likely to replace membrane and mechanical switches in keyboards – compare REED SWITCH

halt *vb* to stop the execution of a program at some point specified by the programmer – **halt** *n* □ see BREAK

hammer *n* a small pivoted flat head which strikes an embossed character against the paper in an impact printer (eg a daisywheel printer) □ see DAISYWHEEL PRINTER

Hamming code *n* a code system used for detecting and correcting errors in a flow of digital information. The Hamming code uses check bits which occupy the '1', '2', '4', etc positions in each binary word. These bits are used as even-parity checks on a different combination of bits in the word. If any bit is changed in transmission then one or more of the check bits will be wrong and the combination of these bits points to the erroneous bit in the word which is then corrected. □ see PARITY CHECK

handshaking *n* the sending and receiving of a predetermined signal between a microprocessor and external equipment indicating when an exchange of data is about to take place or has taken place. For example, an analogue-to-digital converter connected to a microcomputer can produce a signal telling the microprocessor that it has finished converting a temperature measurement into digital form, and that this data is now ready for processing – **handshake** *adj,*

handshake *vb* □ see INTERRUPT

hard copy *n* data (eg a program listing or graphics) printed out on paper – compare SOFT COPY

hard disk *n* a rigid disk of plastic coated with a magnetic film for storing computer data – compare FLOPPY DISK, WINCHESTER DISK

hard-sectored disk *n* a magnetic disk having sectors formed by mechanical means (eg slots cut into the disk hub) – compare SOFT-SECTORED DISK □ see SECTOR

hardware *n* any electronic or mechanical equipment which makes up an electronic system <*microcomputer peripherals are* ~> □ see SOFTWARE, FIRMWARE

hard-wired *adj, of (the components of) a computer system* having its operation determined by the physical way it is put together (wired), and the inputs it receives, rather than on stored instructions <~ *logic circuits*>

harmonic *n* any frequency or a range of frequencies (eg of sound) which is a whole number multiple of a fundamental frequency. In the production of electronic music, harmonics are added to a fundamental frequency in order to 'flavour' it with the characteristics of a particular sound (eg the sound of a harp).

harmonic distortion *n* distortion of an audio signal (eg that produced by an audio amplifier) caused by unwanted harmonics □ see CROSSOVER DISTORTION

Hartley oscillator *n* a transistor circuit used for generating frequencies usu for radio transmitters. The Hartley oscillator makes use of a tapped coil which is part of a tuned circuit in the transistor's collector circuit – compare COLPITTS OSCILLATOR □ see TAP

head *n* that part of a cassette recorder, disk drive, or other recording and playback device which transfers information to and from the recording medium □ see READ/WRITE HEAD

headphones *n* a listening device consisting of a springy band that fits over the head and carries a small speaker for each ear. Headphones for listening to music have two high quality moving coil speakers usu of 8 ohm resistance, and they are sometimes fitted with a separate volume control for each ear. Headphones for listening to speech (eg on a telephone system) usu have a higher resistance and a lower frequency response. □ see EARPHONE

heat sensor *n* – see THERMISTOR, THERMOCOUPLE

heat-shrinkable sleeving *n* hollow plastic tubing which is slipped over bare wire and heated so that it shrinks tightly round the wire to protect it against inadvertent electrical shorts and moisture. Some heat-shrinkable sleeving is made of clear plastic so that soldered joints can be inspected through it, and some is flame resistant.

heat sink *n* a plate or block of metal which is fixed to a semiconductor device in order to prevent the device from being damaged by the heat generated within it. A heat sink is usu painted black and has fins on it in order to make it better able to get rid of the heat. Power transistors in hi-fi equipment and voltage regulators in microcomputers are two devices which generally require a heat sink. □ see THERMAL RUNAWAY

henry *n* (symbol **H**) the unit of electrical inductance. If a potential difference of 1 volt is generated across the terminals of an inductor when the current through it is changing at the rate of 1 ampere per second, the inductance of the inductor is 1 henry. □ see INDUCTOR, INDUCTANCE

hertz *n* (symbol **Hz**) the unit of frequency equal to one complete cycle of an alternating waveform per second <*the frequency of the mains current is 50* ~> □ see FREQUENCY, WAVELENGTH

heterodyning *n* a method used in some radio receiver circuits of changing every incoming carrier frequency to one fixed frequency by mixing the incoming frequency with another high frequency signal to create a 'beat' note. Heterodyning improves circuit stability and sensitivity. – **heterodyne** *vb,* **heterodyne** *adj* □ see BEAT FREQUENCY

heuristic *adj* of or being problem-solving techniques that involve learning from experience, guesses, and trial-and-error <~ *computer programming*> – **heuristically** *adv* □ see ARTIFICIAL INTELLIGENCE, KNOWLEDGE ENGINEERING, FIFTH GENERATION COMPUTER

hex – see HEXADECIMAL

hexadecimal *adj* (abbr **hex**) of or being a number system having a base of 16. A hexadecimal number (eg C6) uses the decimal numbers 0 to 9 and the letters A to F. A computer instruction written in machine code using 8 binary digits can be expressed more simply as a two-digit hexadecimal number. – **hexadecimal** *n* □ see BINARY, DECIMAL, MACHINE CODE

hex dump *n* a group of instructions in a machine code program which are loaded into memory as a single block of data □ see MACHINE CODE

hex pad *n* a simple keyboard for entering hexadecimal numbers into a computer system □ see HEXADECIMAL, KEYBOARD, KEYPAD

hidden line *n* the presentation, on a computer-generated image of an object, of edges and outlines which would normally not be seen when viewing the actual object. Hidden lines are used in computer graphics to give the image perspective and three-dimensional form □ see COMPUTER GRAPHICS

hi-fi *adj* HIGH FIDELITY

high *adj* **1** *of technology* up-to-date and advanced in development and concept <*microelectronics is* ~ *technology*> **2** of or being the larger of the two levels of voltage in a digital logic circuit <high-*level logic is given the binary number 1*> **3** *of the operation of a modem* transmitting data at speeds of up to 9600 bits per second <high-*speed modem*> – compare LOW

high fidelity *also* **hi-fi** *adj* of or being an audio system that produces speech and music which sounds as real as the original <*a* ~ *amplifier*> – **high fidelity** *n* □ see COMPACT DISK, DIGITAL

RECORDING

high-level language *n* a computer language which is user-oriented so that programmers easily understand and learn it *<Basic is a ~>* – compare LOW-LEVEL LANGUAGE □ see COMPILER, MACHINE CODE

highlight *vb* to draw attention to some part of a graphics display on a VDU (eg by flashing it, or using reverse video) – **highlight** *n* □ see BLINK, REVERSE VIDEO

high-pass filter *n* a circuit which stops low frequencies and allows high frequencies to pass through it. One use of a high-pass filter is to cut out the 50 hertz mains hum to prevent it from interfering with the operation of electronic instruments – compare LOW-PASS FILTER □ see BAND PASS FILTER

high resolution graphics *n* images displayed on a VDU which show fine detail. High resolution graphics are produced by controlling the individual phosphor dots (pixels) on the screen. A computer's ability to produce images with good detail depends largely on the amount of screen memory available in the computer. A screen might have 640 phosphor dots along each line. If the screen has 420 lines down it, there is a total of 268 800 dots available for building up images on the screen. If only one colour is displayed, 33 600 bytes (268 800/8) or about 32K of screen memory is required. Very high resolution graphics require very large amounts of screen memory and consequently computers for producing detailed pictures are expensive – compare LOW RESOLUTION □ see GRAPHICS

high-speed *adj* – see SPEED, HIGH 3

highway *n* the main electrical connections between a microprocessor and the memory and peripheral devices making up a computer system □ see BUS

HIRAM – see RAMTOP

hi res *n* [*high res*olution] – see HIGH RESOLUTION GRAPHICS

histogram *n* a graph emphasizing the frequency of occurrence of quantities *<a ~ of monthly rainfall>* □ see GRAPH

hit *vb* to strike a key suddenly *<to ~ the 'space' key in a computer game>*

hit key *n* FIRE BUTTON

hole *n* a vacancy in the crystal structure of a semiconductor which is not occupied by an electron and, therefore, which behaves as if it were a mobile positive charge □ see P-TYPE, VALENCY, ACCEPTOR

hologram *n* an optical (interference) pattern produced on a photographic film when one part of a split beam of laser light meets the other part of the same beam reflected off an object; *also* the three-dimensional picture of this object obtained by illuminating the photographic film with laser light – **holographic** *adj,* **holography** *n* □ see HOLOGRAPHIC MEMORY

holographic memory *n* a memory device in which binary data is stored as an optical (interference) pattern in a photographic film. The holographic memory is not in commercial production but it offers the possibility for storing enormous quantities of data in a very small area or volume which could be read using a laser beam. □ see MAGNETIC BUBBLE MEMORY, OPTICAL MEMORY

home computer *n* a computer for use in the home and generally taken to mean a microcomputer. Home computers are used for playing computer games, handling home accounts, controlling central heating systems, etc. □ see MICROCOMPUTER, PORTABLE MICRO

Hopper, Grace. A pioneer of computing who helped in building valve computers in the USA in the 1940s and in creating software for them. Among her many achievements were three notable contributions: the creation of the high-level language, Cobol; the creation of the first compiler to help translate high-level languages into machine code; and the discovery of the first 'bug' – a moth had been hammered to death in the contacts of a relay in the Harvard Mark 2 computer. She successfully 'debugged' the computer with tweezers and the insect is glued beside the entry for 15.45 on 9 September 1945 in the computer's log book kept at the Naval Museum in Virginia! □ see BUG, COBOL

host computer *n* **1** a computer which has overall control in a communications network (eg Prestel) **2** a computer that prepares programs for execution on another type of computer

host processor *n* a microprocessor which has overall control of other microprocessors in a multiprocessor system □ see PARALLEL PROCESSING

housekeeping *n* the routine tasks that a computer has to do if it is to work properly. Housekeeping includes keeping track of the amount of computer memory being used by a program and ensuring that memory space is used efficiently

hum *n* – see MAINS HUM

hunting *n* an unwanted low frequency oscillation in an electronic control system caused by too much feedback □ see FEEDBACK

hybrid computer *n* a computer which handles both analogue and digital signal processing

hybrid integrated circuit *n* an integrated circuit which incorporates a conventional monolithic integrated circuit linked to discrete devices such as capacitors and resistors, all formed as a thin film on a silicon substrate

hysteresis *n* a delay between cause and effect (eg the output state of a logic circuit lagging behind a change in input state). Hysteresis is used in the design of Schmitt triggers. □ see SCHMITT TRIGGER

Hz the symbol for hertz □ see HERTZ, FREQUENCY

I

I the symbol for electric current □ see ELECTRICITY, CURRENT
IC – see INTEGRATED CIRCUIT
icon *n* a symbolic representation on a VDU of a facility available to the user of a computer system. Thus a business computer might display an icon of a rubbish bin so that any piece of work (eg the result of a calculation performed by another icon of a pocket calculator) can be moved on screen to the rubbish bin, using an input device such as a mouse, and disposed of. □ see MOUSE, BUSINESS COMPUTER
IEEE [*I*nstitute of *E*lectrical and *E*lectronic *E*ngineers] – see IEEE-488 INTERFACE
IEEE-488 interface *n* also **I-triple-E interface** a standardized method for connecting computers to peripheral equipment such as plotters, disk drives, signal generators, and test equipment. The IEEE-488 interface uses eight lines to send data in parallel, and eight lines for control signals. The maximum length of the lines is about 20 metres and uses a 25-pin D-type connector or a DIN plug – compare RS-232 INTERFACE □ see CENTRONICS INTERFACE, PARALLEL INTERFACE
IGFET [*i*nsulated-*g*ate *f*ield-*e*ffect *t*ransistor] – see FIELD-EFFECT TRANSISTOR
I²L – see INTEGRATED INJECTION LOGIC
image *n* a likeness to an object or person created electronically <*a VDU produces ~s*> □ see TELEVISION, ULTRASONIC SCANNER, ELECTRON MICROSCOPE
image analysis *n* computer techniques for breaking down an image into component parts in order to identify it. Image analysis (and speech recognition) are two key areas that are being developed to make the interfacing of people and machines easier. □ see SPEECH RECOGNITION, ARTIFICIAL INTELLIGENCE, ROBOTICS, IMAGE RECOGNITION
image converter *n* a device for changing an invisible image (eg one produced by infrared light) into one which is visible □ see NIGHT SIGHT, CHARGE-COUPLED DEVICE
image recognition *n* the identification of an object by using computers linked to TV cameras and/or other light sensors. Image recognition is being developed so that robots can make selections from a 'mixed bag' of components on a production line. □ see ROBOT, ARTIFICIAL INTELLIGENCE
image sensor *n* any device for detecting an image and converting it into a pattern of electrical signals. □ see CHARGE-COUPLED DEVICE, VIDEO CAMERA
immediate addressing *n* a method of telling a computer, using machine code, where to find the next piece of data by letting the operand itself contain a piece of data. The op code then forms part

of the machine code instruction that contains the name of a register.
□ see OP CODE, OPERAND, DIRECT ADDRESSING, INDIRECT
ADDRESSING

impact matrix printer *n* a printer which forms characters by striking
small rods (stylii) against ribbon or paper □ see DOT MATRIX
PRINTER, THERMAL MATRIX PRINTER

impedance *n* (symbol **Z**) the resistance of a circuit to alternating
current (eg the mains current) flowing through it. Impedance is due
to the combined effects of resistors and capacitors and/or inductors.
It varies with the frequency of the alternating current and is
measured in ohms – compare REACTANCE

impedance matching *n* a method of ensuring that signals are
efficiently transferred from one circuit to another by making the
output impedance of one circuit equal to the input impedance of the
other □ see IMPEDANCE, EFFICIENCY

implementation *n* the designing, making, testing, and installing of a
program or computer system □ see TURNKEY

impurity *n* foreign material (eg boron) added to a pure
semiconductor (eg silicon) to produce an intrinsic semiconductor (eg
p-type) with properties useful for making transistors and integrated
circuits □ see SEMICONDUCTOR, DOPE, N-TYPE, P-TYPE, VALENCY

increment *vb* to increase the value of a variable usu by one or the
value of the contents of a register or memory location – compare
DECREMENT – **increment** *n*

indexed addressing *n* a method of telling a computer, using
machine code, where to find the next piece of data by letting the
operand be only part (called the base address) of the address
needed. To complete the address, the contents (called the
displacement value) of a register called the index register must be
added to it. □ see IMMEDIATE ADDRESSING, INDIRECT ADDRESSING

index register *n* a special register in a microprocessor used for
working out the next piece of data to be used in a calculation □ see
REGISTER, INDEXED ADDRESSING

indicator *n* any device which shows the condition of something (eg a
red light might show that power is switched on to an amplifier) –
indicator *adj* □ see LIGHT-EMITTING DIODE, CONSOLE

indirect addressing *n* a method of telling a computer, using
machine code, where to find the next piece of data by letting the
operand be the address of the memory location or register where the
address of the data is stored. The address given in the operand is
also known as the action address, dispatch address, or jump vector.
□ see IMMEDIATE ADDRESSING, DIRECT ADDRESSING, OPERAND

inductance *n* (symbol **L**) the property of a circuit which makes it
generate a voltage when a current either in the circuit itself or in a
nearby circuit changes □ see HENRY, INDUCTOR, TRANSFORMER,
LENZ'S LAW

inductive reactance *n* – see REACTANCE

inductor *n* also **choke** an electrical component designed to resist the
flow of a changing current through it. An inductor is widely used for
protecting circuits against sudden surges of voltage, and, in
combination with a capacitor, to produce a tuned circuit for a radio
receiver. □ see HENRY, TUNED CIRCUIT, TRANSFORMER

industrial robot *n* – see ROBOT, AUTOMATION, ARTIFICIAL
INTELLIGENCE

information processing *n* DATA PROCESSING

information provider *n* (abbr **IP**) a person or organization that
produces pages of information for display on videotext terminals.
British Telecom's Prestel service provides over 200 thousand pages
of information produced by a large number of information providers
(eg travel agents). □ see VIDEOTEXT, PRESTEL

information retrieval *n* the cataloguing and arranging of
information (eg library services) for access by users □ see PRESTEL,
DATABASE

information society *n* a society of the foreseeable future which is
largely dependent on microelectronics for communicating with
people far and wide □ see INFORMATION TECHNOLOGY

information technology *n* (abbr **IT**) the gathering, processing, and
circulation of information by combining the data-processing power of
the computer with the message-sending capability of
telecommunications. Microelectronics is providing the major
stimulus to the development and use of information technology.
Nowadays, thousands of ordinary people are able to access and
manipulate information from their offices, schools, and homes in a
way previously available to a few specialists in commerce, industry,
and research institutions. It is impossible to predict just how society
will change when everybody has access to powerful computers and
global communications systems. It has been said that the nation that
dominates in the field of information technology will possess the
keys to world leadership in the 21st century. □ see
MICROELECTRONICS, OPTICAL COMMUNICATIONS, COMMUNICATIONS
SATELLITE, ARTIFICIAL INTELLIGENCE, DIRECT BROADCAST
SATELLITE

infrared *adj* of or being electromagnetic radiation having
wavelengths between the red end of the visible spectrum and the
microwave region, ie wavelengths between about 700 nanometres
and 1 millimetre respectively. Infrared rays are invisible to the eye
but they are being increasingly used in electronic systems. For
example, TV and video remote control systems use infrared rays,
and tiny lasers which generate infrared rays are used in the
developing areas of optical communications, the compact disk, and
the video disk. In medicine and architecture electronic sensors can
build up an image of a body or a building showing the presence of
cancer and high heat loss respectively. – **infrared** *n* □ see OPTICAL
COMMUNICATIONS, LASER, COMPACT DISK, VIDEO DISK

ingot *n* BOULE

inhibit *vb* to stop the operation of something <~ing *an electronic counter*> □ see DISABLE

initialize *vb* **1** to give a starting value to a variable in a computer program □ see VARIABLE, STRING **2** to divide a disk (eg a floppy disk) into tracks and sectors which is done by one of the programs in the disk operating system □ see DISK OPERATING SYSTEM, SECTOR

ink-jet printer *n* a device for printing computer programs, data, and graphics characters on paper using a printhead which directs a very fine jet of electrically-charged ink at the paper under the control of the computer

INMARSAT [*In*ternational *Mar*itime *Sat*ellite organization] an organization which administers satellite communications to ships at sea. About 2500 ships are now fitted with a special antenna which always points at the satellite however much the ship heaves and rolls. INMARSAT uses American and European satellites in geostationary orbit and the service is planned to include aircraft to improve the safety and efficiency of airlines. □ see INTELSAT, COMMUNICATIONS SATELLITE

¹**input** *n* **1** the act or process of putting information into a computer or other data processing machine <*a keyboard is an ~ device*> **2** the point (eg a socket) at which information is fed into a machine **3** information fed into a data processing machine <*an ~ to a microcomputer*> – compare OUTPUT

²**input** *vb* to put information into a device

input device *n* a computer peripheral (eg a joystick) which sends data into a computer – compare OUTPUT DEVICE

input/output chip *n* an integrated circuit designed to handle data flowing into and/or out of a data processing system. Thus one purpose of an input/output chip is to convert binary data from parallel to serial form. □ see UART, RS-232 INTERFACE

input/output device *also* **I/O device** *n* a device (eg an integrated circuit) enabling a computer to respond to and act on devices connected to it □ see PERIPHERAL, INPUT/OUTPUT PORT, ANALOGUE-TO-DIGITAL CONVERTER

input/output port *also* **I/O port** *n* an electrical 'window' through which a microcomputer sends data to and receives data from an external device (eg a printer) connected to it. An I/O port consists of an edge connector or other connecting system usu attached to the main printed circuit of the microcomputer. □ see PERIPHERAL, INPUT/OUTPUT CHIP

input sensitivity *n* the minimum amplitude of an input signal at a particular frequency which is required to produce an output signal with a particular signal-to-noise ratio <*an ~ of 5mV*>

instability *n* erratic and unwanted signals produced by a circuit (eg 'howling' or 'motor-boating' sounds from an amplifier)

instruction *n* a written statement (eg Goto) in a computer program, or its equivalent binary representation in machine code, which instructs a computer to carry out a single operation □ see INSTRUCTION SET, LIST

instruction cycle *n* the time taken for each instruction to be fetched from memory, decoded, and executed in the execution of a computer program. An instruction cycle is not constant since each instruction can have from 1 to 3 bytes of binary data – compare MACHINE CYCLE

instruction register *n* a register in a microprocessor that holds an instruction (brought in from memory devices) which is then executed by the control circuits in the microprocessor □ see INSTRUCTION SET, REGISTER

instruction set *n* the complete set of instructions that a computer understands, each instruction setting off one operation (eg Store) inside the microprocessor □ see INSTRUCTION CYCLE

instrumentation *n* the design, construction, and use of devices and systems for making measurements <~ *on a weather satellite*> □ see SENSOR, INTERPLANETARY PROBE

instrument panel *n* that part of equipment on which is mounted switches, meters, and other control and display devices □ see CONSOLE, FRONT PANEL

insulated-gate field-effect transistor *n* (abbr **IGFET**) METAL-OXIDE SEMICONDUCTOR FIELD-EFFECT TRANSISTOR

insulator *n* a material which does not allow electricity, heat, or sound to pass through it <*pvc tape is an electrical* ~> – **insulation** *n*, **insulate** *vb* □ see DIELECTRIC

integer *n* any whole number (eg 6), a set of which make up a number system □ see DECIMAL, BINARY

integrated *adj, of an electronic component* made from many devices all assembled on a single silicon chip <*an ~ circuit*> – compare DISCRETE

integrated circuit *n* (abbr **IC**) an electrical circuit comprising transistors, resistors, diodes, and sometimes capacitors formed and connected together on a single small chip of silicon □ see SILICON CHIP, PHOTOLITHOGRAPHY, MOORE'S LAW, MICROELECTRONICS

integrated injection logic *n* (abbr **I²L**) a family of logic circuits comprising logic gates which have very low power consumption and which are able to be packed very close together on a silicon chip

integrating amplifier *n* a circuit building block based on an operational amplifier which integrates (ie adds up) over a period of time a varying input signal. Integrating amplifiers are used in analogue computers to perform the mathematical operation of integration in the simulation of the behaviour of mechanical systems, and in instrumentation circuits. □ see OPERATIONAL AMPLIFIER

integrator *n* an electronic circuit which produces an average value of a changing input voltage. Integrators are used in some types of digital-to-analogue converters (eg a tachometer).

intelligence *n* the ability of an electronic device or system to learn from experience. In a more restricted sense, a computer terminal is said to have intelligence if it allows the user to edit data before it is sent out. □ see ARTIFICIAL INTELLIGENCE, ROBOTICS, HEURISTIC

intelligent terminal *n* a computer terminal that allows the user to modify data before sending it somewhere else – compare DUMB TERMINAL

INTELSAT [*In*ternational *Tele*communication *Sat*ellite] an organization which masterminds international communications by satellite. At the moment, INTELSAT has 17 operational satellites parked in geostationary orbit; the most recent, the INTELSAT-5 series, can each handle the equivalent of 12 000 telephone calls and 2 colour television channels. INTELSAT are now planning to launch satellites in about 30 years time which will each have a mass of 50 tonnes and be capable of handling 5 million 2-way telephone calls and hundreds of TV programmes □ see COMMUNICATIONS SATELLITE, CABLE TV, GEOSTATIONARY

intensity *n* **1** the strength of a signal *<the ~ of a laser beam>* **2** the brightness of a source of radiation *<the ~ of a light-emitting diode>* □ see AMPLITUDE

interactive *adj* of or being a computer program that allows a person to communicate with the program while it is running. Most adventure games and educational programs allow interactive use through a keyboard, joystick, or light-pen. – **interactively** *adv,* **interaction** *n* □ see ADVENTURE GAME, FIFTH GENERATION COMPUTER

interface *n* a circuit or device which connects a peripheral device to a computer to enable the transfer of data between the two *<a modem is an ~ between a microcomputer and the telephone line>* – **interface** *vb* □ see INPUT/OUTPUT CHIP, INPUT/OUTPUT PORT, PERIPHERAL, UART

interference *n* anything (eg unusual atmospheric conditions) which interferes with the transmission of signals □ see DISTORTION, WHITE NOISE

internal resistance *n* the resistance of a source of electrical power (eg a battery) which is responsible for wasting some of the electrical energy available from the source □ see RESISTANCE

internal storage *n* storage of programs and data within the body of a computer

interplanetary probe *n* a highly sophisticated electronic device sent from Earth to take measurements in space between the planets of the solar system and to observe the planets at close range. Many interplanetary probes have been sent on missions of this kind since the launch of Sputnik in 1957. All the planets, except the outer planets of Neptune and Pluto, have so far been visited by probes, some of which have landed on the surface of the planets to make observations on the spot. The Moon, of course, has been at the centre of investigation by interplanetary probes for more than 20 years. Even the Sun at the centre of the solar system has been examined at close range from within the orbit of Mercury. In addition, special missions to comets and asteroids are being planned. None of these missions to the far reaches of the solar system would have been possible without the remarkable advances in microelectronics over the last 25 years, esp its use in the development of sophisticated communications, computing, and instrumentation systems. □ see VOYAGER, VIKING, GALILEO, IRAS, INTELSAT, SPACE SHUTTLE, COMMUNICATIONS SATELLITE

interpreter *n* a program in a computer which converts a high-level language (eg Basic) into machine code one instruction at a time – compare COMPILER

interrogate *vb* to get data from a storage system *<to ~ a database>* □ see ACCESS

interrupt *n* a process started by a computer peripheral which interrupts a program in progress and causes the computer to execute a special program associated with the interrupt and then return to the previous program as though no interruption had taken place – **interrupt** *vb,* **interruption** *n* □ see HANDSHAKING

intrinsic *adj, of current flow through pure silicon or germanium* produced by the action of heat breaking the bonds between atoms to give equal numbers of electrons and holes – compare EXTRINSIC □ see SEMICONDUCTOR

intruder detector *n* – see BURGLAR ALARM, SECURITY ELECTRONICS

inverse video *n* REVERSE VIDEO

inverter *n* a device for changing a digital signal from a value of 0 to a value of 1, and vice versa. Integrated circuits containing several inverters are useful when interfacing equipment to microcomputers

I/O [*input/output*] – see INPUT/OUTPUT DEVICE, INPUT/OUTPUT PORT

ion *n* an atom or group of atoms which has gained or lost one or more electrons and which therefore carries a positive or negative electrical charge □ see IONIZATION, CHARGE, COULOMB

ion implantation *n* the process of producing microscopically small p-type or n-type regions in a pure silicon wafer by heating in a furnace containing a dopant (eg boron) in the form of electrically charged particles (ions). The doped regions are the building blocks of transistors making up the circuit on a silicon chip. □ see DOPANT, P-TYPE, N-TYPE, SILICON CHIP, PHOTOLITHOGRAPHY

ionization *n* the separation of one or more electrons from their parent atoms to create free electrons and positively charged atoms. Ionization is responsible for the emission of light in some types of display (eg planar gas discharge displays), and in fluorescent tubes. The electrons which leave the cathode of a cathode-ray tube momentarily leave behind atoms which have lost electrons – **ionize** *vb*

ionosphere *n* the portion of the Earth's atmosphere which enables radio communication round the Earth for waves below a frequency of about 30 megahertz. The ionosphere stretches from about 50 kilometres to about 500 kilometres above the Earth's surface and contains positively charged ions generated by the impact of the Sun's ultraviolet radiation on air molecules in the Earth's upper atmosphere. □ see ION, SKY WAVE

IP – see INFORMATION PROVIDER

IRAS [*I*nfrared *A*stronomy *S*atellite] a satellite, refrigerated to about −257C, whose sensors can detect the faint radiation from clouds of dust in interstellar space where new stars are being born. IRAS has also discovered new comets. □ see INFRARED, CRYOGENIC, ABSOLUTE TEMPERATURE SCALE

isolator *n* a device which electrically insulates two parts of a circuit but allows signals to pass between them. An isolator (eg an optoisolator) is generally used to make sure that any high power and/or high voltage devices which a circuit controls cannot damage the controlling circuit should a fault occur. □ see OPTOISOLATOR, TRANSFORMER, RELAY, JUNCTION ISOLATION, OXIDE ISOLATION

IT – see INFORMATION TECHNOLOGY

iterative *adj, of a calculation* getting a progressively more accurate result by repeating the calculation a number of times <*an ~ calculation on a programmable calculator*> – **iterate** *vb,* **iteratively** *adv*

J

J

J the symbol for joule □ see JOULE

jack plug *n* a connector on the end of a lead attached to devices such as headphones, power supplies, and modems. A jack plug has two or more metal bands (poles) round a cylindrical pin which make contact with springy connectors when plugged into a jack socket.

jack socket *n* – see JACK PLUG

jitter *n* unwanted and irregular back-and-forth movement of an image displayed on a VDU or cathode-ray tube – compare JUDDER

JK flip-flop *n* an integrated circuit device used in binary counters and other digital circuit applications. The JK flip-flop has two inputs called J and K (for no obvious reason), and the binary signals applied to these inputs determines whether the flip-flop acts as a T-type flip-flop or a D-type flip-flop. □ see T-TYPE FLIP-FLOP, D-TYPE FLIP-FLOP

job *n* a number of tasks making up a unit of work for processing by a computer

Josephson Effect *n* the 'tunnelling' of electrons through a thin layer of electrically insulating material separating two superconductors □ see JOSEPHSON MEMORY, SUPERCONDUCTOR, CRYOGENICS

Josephson junction *n* the transition between two superconductors separated by an electrical insulator, which might be harnessed in the memories of future fast computers. In a Josephson junction electrons are able to 'tunnel' through the junction when the strength of a local magnetic field changes the superconductor from its normal to its superconducting state. In order for the superconductor to be in a superconducting state it must be cooled to the temperature of liquid helium, ie within a few degrees of absolute zero. The Josephson junction is faster and operates at lower power than present-day semiconductor memories, therefore memories using it will be very fast in operation and have a very high component density. □ see JOSEPHSON MEMORY, SUPERCONDUCTORS, CRYOGENIC MEMORY, ACCESS TIME

Josephson memory *n* a high-speed, low-power, and nonvolatile memory device which makes use of the Josephson Effect and which could be used in future computer systems – compare MAGNETIC BUBBLE MEMORY

joule *n* a unit for measuring energy. The joule is defined as the work done when a force of 1 newton (N) moves its point of application a distance of 1 metre (m). In electronics, the joule is used mainly for measuring electrical and light energy. □ see POWER, WATT

joystick *n* a hand-operated lever mounted in a control box and connected to a microcomputer to move graphics characters about the screen of a VDU. A joystick usu includes a 'fire' button for controlling action in a computer game. □ see MOUSE, TRACK BALL

ROLLER

judder *n* unwanted and jerky vibrations of an image displayed on a
VDU in which parts of the image overlap temporarily – compare
JITTER

JUGFET [*ju*nction *g*ate *f*ield-*ef*fect *t*ransistor] – see FIELD-EFFECT
TRANSISTOR

jump *n* an instruction in machine code which causes a computer to
stop executing a program and move to another part of the program

jump lead *n* also **jumper** a (usu short) length of wire enabling
connections to be made between two parts of a circuit or between
two separate circuits □ see BREADBOARD

junction *n* a region of electrical contact between two dissimilar
metals (eg a thermocouple junction) or between two dissimilar
semiconductors (eg a pn junction) which has useful electrical
properties □ see THERMOCOUPLE, PN JUNCTION

junction capacitor *n* a capacitor formed on a silicon chip from a
reverse-biased pn junction in the manufacture of integrated circuits.
This type of capacitor gives only a small capacitance up to a
maximum value of about 100 picofarads, unlike resistors which can
be made from semiconducting materials to have large values of
resistance – compare MOS CAPACITOR □ see PN JUNCTION, VARICAP
DIODE

junction diode *n* a diode formed by creating a p-type semiconductor
in a thin slice of n-type semiconductor using the technique of gaseous
diffusion – compare POINT CONTACT DIODE

junction isolation *n* a way of providing electrical insulation between
two or more components on a silicon chip by using the high
resistance properties of a reverse-biased pn junction. Junction
isolation enables transistors and other devices, formed on a silicon
chip during the manufacture of an integrated circuit, to be
electrically isolated from each other – compare OXIDE ISOLATION
□ see PHOTOLITHOGRAPHY, ETCHING

K

k the symbol for the prefix kilo meaning one thousand times □ see
KILOMETRE

K 1 a symbol for 1024 **2** a symbol for a memory capacity of 1024
bytes □ see KILOBYTE

Karnaugh map *n* a type of truth table enabling designers to
understand the operation of complex logic circuits (eg multiplexers)
□ see TRUTH TABLE, VENN DIAGRAM, FLOWCHART

key *n* a switch which is pressed or touched to enter data into a
system <*a ~ on an electronic typewriter*> – **key** *vb* □ see
KEYBOARD, KEYPAD

keyboard *n* a group of keys arranged in a certain pattern, by which
information is entered into an electronic system □ see QWERTY
KEYBOARD, KEYPAD

keyboard encoder *n* the digital circuit inside a computer or
calculator which scans the keys of a keyboard rapidly to detect when
a particular key is pressed. This information is then converted into
binary code and sent to the microprocessor circuits as data or
instructions. □ see KEYBOARD, ENCODER

key in *vb* to enter data or instructions via a keyboard or keypad <*to
~ one's bank account number*>

keypad *n* a small keyboard with a restricted range of keys <*a ~ on
a numerically controlled machine tool*> □ see KEYBOARD

keyword *n* a word (eg PRINT in Basic) that is entered at a
keyboard and which a microcomputer has been programmed to
recognize and respond to in a particular way □ see FUNCTION KEY

Kilby, Jack. Inventor of the integrated circuit (while working at
Texas Instruments, USA, in 1958). His device was about 12×6
millimetres and contained a few transistors. □ see MOORE'S LAW,
INTEGRATED CIRCUIT

kilobit *n* a unit for measuring binary data equal to either 1000 or
1024 bits <*the capacity of a high volume digital communications
channel is 9.6 ~s per second*> □ see KILOBYTE

kilobyte *n* (symbol **KB**) 1000 or 1024 bytes of computer data □ see
BYTE, MEGABYTE

kilohertz *n* (symbol **kHz**) a frequency equal to 1000 (10^3) hertz
□ see LONG WAVES, FREQUENCY

kilohm *n* (symbol **kΩ**) an electrical resistance equal to 1000 (10^3)
ohms □ see RESISTANCE

kilometre *n* (**km**) a distance equal to 1000 (10^3) metres

Kirchoff's laws *n* two laws frequently used to work out the patterns
of current flow and potential difference in networks of resistors
□ see OHM'S LAW

kitchen electronics *n* the use of electronic devices, esp the
microprocessor, to improve the performance, ease of use, and

efficiency of kitchen appliances (eg cookers, washing machines, toasters, dishwashers, food mixers, and refrigerators). The different functions of an appliance are held in a read-only memory, and a microprocessor provides the timing and control sequences. A washing machine, for example, uses microelectronic devices to carry out standardized wash cycles, while at the same time incorporating maximum flexibility for manual override by the user when necessary (eg for dispensing detergents, bleaches, etc). Most appliances incorporate sensors (eg for temperature and water level) which give information to the microprocessor. Control signals operate relays, triacs, and solenoids to switch power on and off to motors, heaters, and water pumps. The microwave oven is one appliance at the centre of this new technology; some of them can provide spoken information about cooking specific foods. On pressing touch controls, the 'chef' can expect that the built-in programmable timers will cook the food properly. □ see ELECTRONIC SEWING MACHINE, SPEECH SYNTHESIS, TRIP COMPUTER, ELECTRONIC ENGINE MANAGEMENT

kludge *n* a badly designed program or system which works but which might give some trouble later □ see BUG

knowledge engineering *n* a study of the use of knowledge in solving problems. Knowledge engineering is of considerable importance in the development of machines capable of intelligent behaviour. □ see KNOWLEDGE SYSTEM, ARTIFICIAL INTELLIGENCE, EXPERT SYSTEM, FIFTH GENERATION COMPUTER

knowledge system *n* a computer-based system providing rapid access to information □ see EXPERT SYSTEM, DATABASE, KNOWLEDGE ENGINEERING

L

L the symbol for inductance □ see INDUCTANCE

lamp *n* a device which gives off light and is used as a source of illumination or as an indicator. A light-emitting diode is sometimes called a 'solid-state lamp'. □ see INDICATOR, LIGHT-EMITTING DIODE

LAN – see LOCAL AREA NETWORK

Landsat any of a series of artificial satellites (eg Landsat 4 launched in 1982) carrying sensors (eg infrared cameras) to look for crop diseases, search for new mineral resources, and keep a check on pollution (eg oil spillage at sea). Earth stations process the digital information from Landsat and produce detailed colour pictures of the Earth's surface. The colours are not true but show heathland in red-brown, corn at harvest time in green, towns and ploughed fields in blue-grey, clear water in deep blue, polluted water in light blue, and so on. □ see ARTIFICIAL SATELLITE, INTELSAT, INMARSAT

language *n* a set of symbols, words, and rules used to write a program of instructions for a computer <*Basic is a high-level computer* ~> □ see SYNTAX, HIGH-LEVEL LANGUAGE, LOW-LEVEL LANGUAGE

large scale integration *n* (abbr **LSI**) the process or technology of making integrated circuits with between 100 and 5000 logic gates on the chip □ see MOORE'S LAW

laser *n* [*l*ight *a*mplification by the *s*timulated *e*mission of *r*adiation] an optical device which produces an intense narrow beam of light or infrared radiation over a narrow frequency band. The gallium arsenide laser is used to produce infrared light for carrying information along optical fibres. Lasers are used in surgery (eg for fixing a detached retina), for reading information on video disks, and for purely artistic effects, and they are being developed for military use (eg for destroying enemy satellites). □ see VIDEO DISK, OPTICAL COMMUNICATIONS, COMPACT DISK, LASER DIODE

laser diode *n* a compact solid-state device (as opposed to a gas-filled laser) which produces a narrow beam of light all of one wavelength (or frequency). The light from a laser diode is used in optical communications systems as the carrier wave for messages. □ see OPTICAL COMMUNICATIONS, CARRIER WAVE, COHERENT LIGHT

laser printer *n* a high-speed printer which uses a fine beam of laser light to write characters formed from a matrix of dots on selenium coated paper. Many modern laser printers use a laser diode to generate the laser beam and the printing speed can be as high as 28 pages per minute with a dot density of 14 thousand dots per square centimetre. Laser printers are extremely versatile and provide a wide range of fonts. Users are able to design graphics, intermix fonts, and print reports which have been merged with graphics and data. However, laser printers are considerably more expensive than

conventional dot matrix printers and daisywheel printers. □ see
PRINTER, LASER

last in first out – see LIFO

latch *n* a circuit for catching and holding onto data until it is needed.
Integrated circuits containing several latches are available and these
are often used in designing peripheral devices to interface with
computers – **latch** *vb*

layout *n* the plan or design of a circuit or system *<the ~ of
components on a printed circuit board>*

LCD – see LIQUID CRYSTAL DISPLAY

LDR – see LIGHT-DEPENDENT RESISTOR

lead *n* a wire or cable for connecting devices together, esp for
connecting a device to a power supply *<a power ~>*

leader *n* the blank uncoated section of tape at the beginning of a
magnetic tape □ see MAGNETIC TAPE

leading edge *n* **1** the first part of a signal waveform during which its
amplitude is rising **2** the edge of a punched card which is the first to
enter a card reader – compare TRAILING EDGE

leading zero suppression *n* (abbr **LZS**) the ability of a calculator
or computer to prevent the display of zeros in front of the first digit
of a number. For example, an 8-digit display with leading zero
suppression would show 123.45 not 000123.45.

leakage current *n* a usu small flow of unwanted current between
electrical conductors in a circuit, esp the steady current flowing
between the terminals of a capacitor

least significant bit *n* (abbr **LSB**) the rightmost binary digit in a
binary word *<the ~ in the word 10001101 is 1>* – compare MOST
SIGNIFICANT BIT □ see BINARY NUMBER, WORD

LED – see LIGHT-EMITTING DIODE

lens *n* **1** a curved piece of glass for focussing visible light or infrared
<a ~ in a video camera> **2** a set of coils and/or electrodes for
focussing a beam of electrons *<a magnetic ~ in an electron
miscroscope>*

Lenz's Law *n* a law stating that when a current flows through a coil
of wire, a magnetic field is set up which tends to oppose the change
of current producing the field. Lenz's law helps in understanding the
operation of inductors, transformers, relays, and other
electromagnetic devices. □ see ELECTROMAGNETIC INDUCTION,
INDUCTOR, TRANSFORMER, RELAY, REACTANCE

level *n* – see HIGH, LOW

library *n* a collection of pictures, pages, charts, etc stored in a
computer system and accessed from a keyboard □ see DATABASE

LIFO [*l*ast *i*n *f*irst *o*ut] a method of storing and retrieving information
in a memory such that the last piece of data stored is the first one to
be retrieved – compare FIFO

light-dependent resistor *n* (abbr **LDR**) a semiconductor device
having an electrical resistance which decreases with the amount of

light falling on it. Light-dependent resistors are usu made from the material cadmium sulphide; they are used in cameras and in a wide range of automatic devices controlled by light. □ see PHOTOTRANSISTOR, SOLAR CELL, OPTOELECTRONICS, SENSOR

light-emitting diode *n* (abbr **LED**) a small semiconductor diode which glows when current passes through it from its anode to its cathode terminal. LEDs are used in many types of electronics display. They can be made to emit light of any colour (red, green, and yellow are common) by careful choice of the impurity added to the base semiconductor (eg gallium phosphide or gallium arsenide). Infrared-emitting LEDs are used widely in optical communications systems, esp for the remote control of domestic TVs – compare LIQUID-CRYSTAL DISPLAY □ see DIODE, SEVEN-SEGMENT DISPLAY, DOT-MATRIX DISPLAY, LASER DIODE

light-pen *n* also **wand** a penlike device connected to a graphics terminal that enables information on a VDU to be drawn and modified. The computer is programmed to recognize the position of the pen on the screen and then to erase information, draw new information, or 'drag' shapes around the screen. □ see GRAPHICS TERMINAL, COMPUTER-AIDED DESIGN, MOUSE, BAR CODING

light sensor *n* – see PHOTOCELL, SOLAR CELL, PHOTODIODE

light waves *n* electromagnetic radiation in the visible spectrum. Light waves are increasingly being used to carry information. □ see OPTICAL COMMUNICATIONS, LASER, OPTOELECTRONICS

line *n* a wire connection carrying information between two or more points <*a telephone* ~>

linear *adj* producing an output signal which varies smoothly with the input signal <*an audio amplifier is a* ~ *device*> □ see AMPLITUDE, GAIN

line density *n* the number of lines in a vertical distance (usu 10 millimetres or 1 inch) that can be displayed on a VDU or printed by a printer

line frequency *n* the frequency needed in a television for the scanning of the electron beam which builds up the picture <*a 625 line TV needs a* ~ *of 15 625 hertz*> □ see FREQUENCY, RASTER SCANNING, TELEVISION

line number *n* a decimal number at the beginning of a line of instructions in a program which identifies that line for future use □ see LISTING

¹link *vb* to make a connection between two parts of a system <*to* ~ *computers together in a network*> □ see NETWORK

²link *n* **1** a communications channel <*a satellite* ~> **2** a short wire connection on a circuit board

liquid crystal display *n* (abbr **LCD**) a display used in a wide range of digital devices (eg calculators, pocket games, and watches) which operates by reflected light rather than producing light as in the light-emitting diode (LED). A liquid crystal is an organic carbon compound which behaves as if it were both a solid and a liquid, ie its molecules readily take up a regular pattern as in a crystal and yet it flows as a liquid. In the liquid crystal display, this compound is sandwiched between two closely spaced transparent metal electrodes which are in the form of the pattern (eg a seven-segmented digit) which is to be displayed. When a voltage is applied across selected electrodes, the molecular arrangement within the liquid crystal changes and shows up as a dark area between the electrodes. Liquid crystal displays use much less current than light-emitting diodes so they are used in all types of portable battery-operated equipment (eg watches); however, they are more expensive and cannot be seen in the dark without illumination – compare LIGHT-EMITTING DIODE

Lisp *n* [*list p*rocessing] a computer language developed in the 1960s and widely used since in the field of artificial intelligence which involves continually searching and comparing lists of data □ see ARTIFICIAL INTELLIGENCE

[1]list *vb* to print out a set of program instructions or data

[2]list *n* items of data or instructions arranged in a (vertical) order *<a program ~>*

listing *n* the complete set of instructions in a computer program which is displayed on the screen of a VDU or printed out on paper

lithium cell *n* a primary cell (ie not rechargeable) which has a long shelf life, can operate at low temperatures, and is a compact source of electrical energy. A lithium cell provides a voltage across its terminals of 3 volts. It is widely used in digital watches, hearing aids, memory cards, microphones, toys, calculators, and similar small-size electronic devices. One recent lithium cell is only 2 millimetres in diameter and 11 millimetres long, and is designed to fit inside a float to illuminate a light-emitting diode for night-time fishing.

liveware *n* a rather facetious reference to the place of people in the design and operation of computer systems □ see HARDWARE, SOFTWARE, FIRMWARE, WETWARE

[1]load *vb* **1** to put program data into a computer *<~ing a games program>* **2** to make an electrical device deliver power to another device *<the loudspeaker ~s the amplifier>*

[2]load *n* **1** a device (eg a motor) which absorbs electrical power from a battery or circuit **2** the electrical power drawn from a circuit *<the ~ on the National Grid>* **3** the amount of communications traffic on a channel □ see OVERLOAD

loader *n* a program which accepts a program in machine code and places it in memory before it is executed □ see COMPILER

load life *n* the number of hours that a cell or battery can deliver electrical current at a specified rate

load line *n* a straight line drawn on a graph of the output characteristics of a transistor amplifier, which is used by a circuit designer to find out the range of signal strengths the amplifier can handle □ see CHARACTERISTIC

local area network *n* (abbr **LAN**) a communications system (eg Ethernet) linking together computer terminals in an office building or manufacturing plant. One use of a local area network is the replacement of the written memorandum by an electronic mailbox facility. □ see ETHERNET, ELECTRONIC OFFICE, PRESTEL

locate *vb* **1** to find out where data is stored in a computer's memory □ see ADDRESS **2** to work out the source of a problem *<to ~ a hardware fault>*

location *n* a place in computer memory, on a disk, etc where data is stored until it is required *<to address a memory ~>*

log *n* a record of a sequence of tasks *<a ~ of computer jobs>* – **log** *vb*

logger *n* DATA LOGGER

logic *n* **1** the use of Boolean algebra, truth tables, or similar methods to work out how digital circuits respond to binary signals *<computer ~>* **2** circuits which respond in a predictable way to binary signals *<combinational ~>* □ see BOOLEAN ALGEBRA, COMBINATIONAL LOGIC, SEQUENTIAL LOGIC

logic analyser *n* an instrument for testing the function of digital circuits. A logic analyser displays the logic signals produced by a circuit on an oscilloscope or on one or more light-emitting diodes □ see LOGIC PROBE

logic array *n* – see PROGRAMMABLE LOGIC ARRAY

logic diagram *n* a circuit diagram showing how logic gates and other digital devices are connected together. A logic diagram uses a set of accepted symbols to represent logic gates, flip-flops, etc *<a ~ of a digital watch>*

logic family *n* either of the two main types of digital integrated circuits used by circuit designers today. One logic family is called transistor-transistor logic (TTL) and the other is called complementary metal-oxide semiconductor logic (CMOSL). □ see TRANSISTOR-TRANSISTOR LOGIC, COMPLEMENTARY METAL-OXIDE SEMICONDUCTOR LOGIC

logic function *n* a function that expresses the operation of a building block (eg an AND gate) in a digital circuit which produces a binary output (1 or 0) determined by the combination of binary signals on its inputs □ see GATE 1

logic gate *n* GATE 1

logic levels *n* the two values of voltage, high level and low level, which digital logic circuits produce or respond to, and which are given the binary numbers 1 and 0. For a particular logic family (eg TTL), there is a range of voltages which qualifies as 'low' or 'high'. □ see LOGIC FAMILY

logic probe *n* a handheld penlike instrument for testing the logic states (1 or 0) of signals in digital circuits. A logic probe has a probe at one end with built-in light-emitting diodes to indicate the presence or absence of a signal. It may be powered by batteries, or it may pick

up its power supply from the circuit under test. □ see LOGIC
ANALYSER

logic symbol *n* a character or shape which represents a logic
function in logic diagrams <*& is the ~ for AND*>

log in *vb* LOG ON

Logo *n* a high-level computer language invented by Seymour Papert
to enable young children to learn about mathematics and spatial
relationships 'painlessly', through computer control of a moving
graphics character or a mobile vehicle called a turtle. □ see TURTLE,
PAPERT, HIGH-LEVEL LANGUAGE

log off *also* **log out** *vb* to end a session at a computer terminal by
entering a command – compare LOG ON

log on *also* **log in** *vb* to begin a session at a computer terminal by
pressing a control key and entering a command and a password –
compare LOG OFF

log out *vb* LOG OFF

long-tailed pair *n* an arrangement of two bipolar transistors used in
the design of differential amplifiers (eg op amps). It gets its name
from the single high-value resistor connected to the two emitters of
the transistors. □ see DIFFERENTIAL AMPLIFIER

long waves *n* radio waves having wavelengths between 1000 and
10 000 metres (ie frequencies of 300 to 30 kilohertz) □ see MEDIUM
WAVES, SHORT WAVES

loop *n* **1** a sequence of instructions in which the last instruction is a
'jump' in machine code or a 'goto' in Basic, so that the sequence is
repeated usu until some variable has reached a preset value <*using a
~ to give a time delay in a program*> **2** an electrical connection in a
circuit which may be 'open' (ie broken) or 'closed' (ie made)

loudspeaker *n* a device for converting electrical energy from an
audio amplifier into sound energy. In most loudspeakers, a copper
coil attached to a paper cone moves in a strong magnetic field when
the electrical signals flow through it – compare MICROPHONE

low *adj* **1** *of technology* not very advanced in development or
concept <*semaphore is ~ technology*> **2** of or being the smaller of
the two levels of voltage in a digital logic circuit <*low-level logic is
given the binary number 0*> **3** *of the operation of a modem*
transmitting data at speeds of less than 600 bits per second <*low-
speed modem*> – compare HIGH

low-level language *n* a computer language designed for
communicating with a microprocessor directly and not readily
understood by humans □ see MACHINE CODE

low-pass filter *n* a circuit which stops high frequencies and allows
low frequencies to pass through it. One use of a low-pass filter is to
cut out high-frequency hiss (eg surface noise from a tape or disk) in a
hi-fi system – compare HIGH-PASS FILTER □ see BAND PASS FILTER

low resolution *adj, of images displayed on a VDU* lacking fine
detail <*Viewdata has ~ graphics*> – compare HIGH RESOLUTION

GRAPHICS
low-speed *adj* – see SPEED, LOW 3
LSB – see LEAST SIGNIFICANT BIT
LSI – see LARGE SCALE INTEGRATION
luminance *n* (the variation in) the degree of brightness of an image (eg a monochrome TV picture) – compare CHROMINANCE
luminesce *vb* to emit 'cold' light <*the screen of a VDU* ~s>
LZS – see LEADING ZERO SUPPRESSION

M

m the symbol for the prefix milli meaning one thousandth of □ see MILLIMETRE

M the symbol for the prefix mega meaning one million times

machine *n* a (computer) device performing some useful and usu complex service – **machinery** *n* □ see MACHINE CODE

machine code *n* instructions in the form of patterns of binary digits which are understood directly by a microprocessor □ see BINARY CODE, LOW-LEVEL LANGUAGE, ASSEMBLY LANGUAGE, OBJECT CODE

machine cycle *n* the time between the 'beats' of the clock which controls the operations performed by a microprocessor. Thus if the clock frequency of a microprocessor is 4 megahertz, the machine cycle is a quarter of a microsecond. □ see CRYSTAL CLOCK, INSTRUCTION CYCLE

machine vision *n* any system comprising sensors (eg TV cameras) and programmable devices which enable a machine, esp a robot, to 'see' and act 'intelligently' □ see ROBOTICS, ARTIFICIAL INTELLIGENCE, IMAGE RECOGNITION

magnet *n* a ferrous material (eg iron and iron oxide) having magnetic North and South poles between which a magnetic field is produced. The magnetic field can be used for exerting forces on other magnets or other materials (eg soft iron in a relay) which become temporarily magnetized by the field. The read/write head in disk drives and cassette recorders acts as a magnet and produces a varying magnetic field to change the magnetic strength of the ferrous material on the disk or tape. □ see FERRITE, READ/WRITE HEAD, MAGNETIC BUBBLE MEMORY

magnetic bubble *n* a minute cylinder-shaped region of magnetism on the surface of a thin crystal of gadolinium gallium garnet which represents a single bit of computer data □ see MAGNETIC BUBBLE MEMORY

magnetic bubble memory *n* (abbr **MBM**) also **bubble memory** a device for storing binary data as a string of magnetic bubbles in a thin film of magnetic material. The magnetic bubbles are created by a current circulating in minute loops in a metallic layer just above the film. A bubble represents binary 1 and its absence, binary 0. The bubbles are made to circulate in a major loop in the film under the control of a rotating magnetic field surrounding the film. A bubble is detected by the changes in resistance of a permalloy strip covering the film. Read, write, and erase stations alongside the major loop enable data to be accessed, stored, and erased, respectively. Data which is to be stored is transferred from the major loop to minor loops until needed when the data is moved back to the major loop. Data to and from major and minor loops is transferred one bit at a time, so a magnetic bubble memory is a serial access device. This

makes it slower in operation than a semiconductor random-access
memory which is a parallel-access device. Nevertheless, for the
expanding area of portable microprocessor products (eg
wordprocessors and microcomputers), a magnetic bubble memory
offers the advantage of high storage capacity in a small volume and,
unlike a random-access memory, it does not lose its data when
power to it is switched off. □ see NONVOLATILE, JOSEPHSON
JUNCTION, CHARGE-COUPLED DEVICE

magnetic disk *n* a disk which has a magnetizable surface on which
computer data can be recorded □ see FLOPPY DISK, HARD DISK

magnetic field *n* the invisible influence round a magnet which
causes one magnet to act on another □ see MAGNET

magnetic ink *n* an ink containing iron oxide used for printing
characters which can be detected and identified by their magnetic
effects □ see MAGNETIC INK CHARACTER RECOGNITION

magnetic ink character recognition *n* (abbr **MICR**) a way of
reading characters printed in magnetic ink on bank cheques and
other documents. The cheque is fed past a read head in which pulses
are generated whenever there are changes in the strength of the
magnetic field produced by the characters. These pulses are analysed
to determine which characters are present on the document. Reading
speeds can be up to 2000 cheques per minute – compare OPTICAL
CHARACTER RECOGNITION

magnetic storage *n* any system (eg magnetic tape) which holds data
in the form of changes in direction of magnetization of a
magnetizable substance □ see MAGNETIC TAPE, MAGNETIC BUBBLE,
MAGNETIC DISK, FERRITE

magnetic tape *n* a strip of material which has a magnetizable
surface on which computer and other data can be recorded

magnetism *n* the attraction of magnetized iron, steel, and some
alloys, and of an electromagnet, for other magnets and magnetizable
substances. Magnetism is the basis of the operation of loudspeakers,
transformers, solenoids, relays, motors, reed relays, and some other
devices used as input and output devices in electronic circuits –
compare ELECTRICITY □ see ELECTROMAGNET, MAGNETIC BUBBLE
MEMORY

magnify *vb* to increase the size of an image displayed on a VDU
□ see ZOOM

magnitude comparator *n* a digital device which compares the value
of two binary numbers and provides a signal when the values are
equal. Magnitude comparators are used for designing computers,
digital instruments, and other digital systems – compare
COMPARATOR

mainframe *n* a powerful computer capable of carrying out many
tasks simultaneously and to which are connected a number of
terminals □ see TIMESHARING

main memory *also* **main store** *n* that part of a computer's memory

(eg random-access memory) containing data which can be directly
accessed by the microprocessor in times as short as a few
nanoseconds – compare BACKING STORE

mains hum *n* the sound sometimes heard from faulty mains-
powered electrical equipment (eg radios and hi fi systems). Mains
hum can be caused by poor smoothing of the rectified power supply
in this equipment, or by leads to speaker units running close to
mains power supply cables. The usual frequency of mains hum is 50
or 100 hertz. □ see RESERVOIR CAPACITOR, FREQUENCY,
INTERFERENCE

majority carrier *n* the more abundant of the two charge carriers,
electrons and holes, in a semiconductor. Electrons are majority
carriers in an n-type semiconductor – compare MINORITY CARRIER
□ see SEMICONDUCTOR, P-TYPE, N-TYPE, VALENCY

malfunction *n* a partial or total failure in the operation of an
electronic device or system – **malfunction** *vb* □ see BUG,
TROUBLESHOOTING

MANIAC [*m*echanical *a*nd *n*umerical *i*ntegrator *a*nd *c*omputer] an
early type of valve computer □ see ENIAC

manipulate *vb* to change data to some more useful form <*an
arithmetic logic unit* ~s *data*>

man-machine interface *n* any hardware which allows a person to
exchange information with a computer or other machine. At present,
a keyboard is the most widely used man-machine interface in
computer systems, but devices are being developed to allow users to
talk to their computers. □ see VOICE CHANNEL, SPEECH SYNTHESIS,
SPEECH RECOGNITION

manual *n* – see DOCUMENTATION

map *n* – see MEMORY MAP, KARNAUGH MAP

mark-to-space ratio *also* **M/S ratio** *n* the ratio of the times a
rectangular wave signal spends its time high (ie 'on') to the time it is
low (ie 'off') □ see ASTABLE

maser *n* [*m*icrowave *a*mplification by *s*timulated *e*mission of
*r*adiation] a device used for amplifying very weak radio signals, esp
those received from satellites and interplanetary probes. A maser
increases the strength of microwaves by energizing atoms to a level
where they give off radio energy of the desired frequency. An Earth
station's dish aerial uses a maser to increase the strength of a weak
radio signal over a million times – compare LASER □ see
MICROWAVES, EARTH STATION

mask *n* **1** a photographic plate used in the manufacture of integrated
circuits □ see PHOTOMASK **2** a bit pattern (eg a byte) which may be
used to select those bits (ie 0s and 1s) which are to be used in a
subsequent computer operation

mask programmable *adj, of a read-only memory* irreversibly
programmed according to a customer's specification by using a metal
mask during manufacture which determines the memory structure

□ see UNCOMMITTED LOGIC ARRAY

mass storage *n* also **bulk storage 1** the storage of large amounts of computer data <~ *on a floppy disk*> **2** backing store capable of storing large amounts of data □ see MAGNETIC BUBBLE MEMORY, MEGABYTE

matching *n* the technique of making a device transfer maximum power or voltage to another device. Matching is achieved in electronic circuits by ensuring that the impedance of the source (eg the secondary winding on a transformer in an audio amplifier) is equal to the impedance of the load (eg the loudspeaker) – **match** *vb* □ see IMPEDANCE, TRANSFORMER

matrix *n* an arrangement of data or devices (eg memory cells) in rows and columns so that each item is identified by two subscripts, its row and column numbers □ see ARRAY

maximum reverse voltage *n* PEAK INVERSE VOLTAGE

MBM – see MAGNETIC BUBBLE MEMORY

medium *n* **1** a material which holds information <*magnetic tape is a data storage* ~> **2** a communication system supplying information to the public <*the* ~ *of television*>

medium scale integration *n* (abbr **MSI**) the process or technology of making integrated circuits with between 20 and 100 logic gates on the chip □ see MOORE'S LAW

medium-speed *adj* of or being data transmission rates between 600 and 4800 bits per second – compare LOW 3, HIGH 3

medium waves *n* radio waves having wavelengths in the range about 200 to about 700 metres (ie frequencies of about 1.5 to about 5 million hertz) – compare LONG WAVES, SHORT WAVES, MICROWAVES

megabit *n* a quantity of binary data equal to 1 million (10^6) bits. Present-day computer memory devices can store this amount of binary data. □ see MEGABYTE

megabyte *n* (symbol **MB**) a quantity of computer data equal to 1 million (10^6) bytes. Present-day floppy disks store this much data.

megahertz *n* (symbol **MHz**) a frequency equal to 1 million (10^6) hertz. Radio waves at this frequency are known as medium waves. A microprocessor's clock may run at between 2 and 4 megahertz. □ see FREQUENCY, CRYSTAL CLOCK, MEDIUM WAVES

megohm *n* (symbol **MΩ**) an electrical resistance equal to 1 million (10^6) ohms □ see RESISTOR

membrane keypad *n* a keyboard comprising switches covered by a flexible plastic film and needing only light pressure to operate them. The membrane is coated on the inside with a thin film of silver alloy and separated by plastic spacers from a lower fixed layer of conducting silver alloy. A membrane keyboard has advantages over a mechanical keyboard in that it can be operated up to 10 million times and it is not affected by moisture and dust; however, the absence of tactile feedback ('feel') makes it unsuitable for rapid operation as required in a wordprocessor, for example. □ see

KEYBOARD, TOUCH-SENSITIVE

memory *n* that part of a computer system holding data and instructions for future use. A typical 8-bit microprocessor can communicate with (address) 65 536 individual memory locations. Some of these locations are used by the computer's read-only memory which is dedicated to essential operations (eg the language and graphics symbols the computer uses). A large proportion of the memory is made available for the computer's random-access memory which acts as a temporary data store for programs entered via the keyboard or loaded into it from a magnetic tape. Some of the remaining locations can be used to communicate with peripherals via the computer's user port. □ see RANDOM-ACCESS MEMORY, READ-ONLY MEMORY, MEMORY MAP

memory map *n* a diagram showing how parts of a computer's memory are allocated to the various functions the computer performs

memory-mapped input/output *also* **memory mapped I/O** *n* a method used by some microcomputers for communicating with devices connected to it, and which involves considering its input/output port as one or more locations in its memory □ see MEMORY, ADDRESS, PERIPHERAL

menu *n* a list of choices available in a computer program. Simple programs have just a few choices in a single list, while complex programs have a main menu from which other menus can be selected

merge *vb* to join two sets of data together to make a single larger set

message *n* any information which travels between a source and destination

metallization *n* the process of making conducting paths between components on an integrated circuit □ see ALUMINIUM, INTEGRATED CIRCUIT

metal-nitride-oxide-semiconductor *n* (abbr **MNOS**) a type of electrically-alterable read-only memory (EAROM) which can be programmed by means of electrostatic charges in a film of silicon nitride. MNOS can be reprogrammed while in a printed circuit board. □ see READ-ONLY MEMORY, FAMOS, FIELD-EFFECT TRANSISTOR

metal-oxide semiconductor field-effect transistor *n* (abbr **MOSFET**) *also* **insulated-gate field-effect transistor** (abbr **IGFET**) a type of field-effect transistor available in two forms: as an NMOS device, the current is largely an electron flow, and as a PMOS device the current is largely a flow of holes □ see FIELD-EFFECT TRANSISTOR, NMOS, PMOS

meter *n* a device which displays information in analogue or digital form □ see MULTIMETER

MICR – see MAGNETIC INK CHARACTER RECOGNITION

micro *n* **1** MICROPROCESSOR **2** MICROCOMPUTER

microchip *n* – see SILICON CHIP

microcircuit *n* a miniature circuit assembled on a tiny piece of pure silicon □ see SILICON CHIP, INTEGRATED CIRCUIT, MOORE'S LAW

microcomputer *also* **micro** *n* a portable desk- or table-top digital computer which can be programmed to perform a wide variety of useful functions quickly and cheaply. Its operations are controlled internally by a microprocessor and other microelectronic devices. Programs can be entered into a microcomputer manually through a keyboard, or loaded into it from an external data store (eg a magnetic tape). The program instructions and the results of its computations can be seen on a VDU. Most microcomputers for home use (home computers) are used to play computer games, but they can also be used to run educational programs, to act as filing and accounting systems, to provide the services of a home doctor, to control a central-heating system, and, via the telephone, to access databases and other sources of information. In hospitals, microcomputers are used to monitor the body functions of critically ill patients and to process and display information from diagnostic equipment (eg ultrasonic scanners). In the manufacturing industries, microcomputers control machines for cutting, shaping, welding, and painting metals and for testing, measuring, and analysing a wide range of processes. While most microcomputers are at present based on 8-bit microprocessors, 16-bit microcomputers which provide much greater computing power and improved graphics are becoming popular. The versatility of microcomputers is enhanced by an increasingly wide range of peripherals such as printers, modems, digitizers, speech processors, and synthesizers. □ see MICROPROCESSOR, RANDOM-ACCESS MEMORY, READ-ONLY MEMORY, MAGNETIC TAPE, MAGNETIC DISK, INFORMATION TECHNOLOGY

microdrive *n 1* a small disk drive for reading and writing from microfloppy disks *2* a small unit containing a fast-running magnetic tape for storing a large amount of data □ see MICROFLOPPY DISK, STRINGY FLOPPY

microelectronics *n* the production and use of silicon chips on which complex circuits containing many thousands of components have been formed. Microelectronics is having a great influence on our work and leisure. It is not that the silicon chip is just another piece of electronic gadgetry which enables new products to be built, but it is the power these products have to produce social change. These changes are already being felt in industry and commerce as robots replace people, and as more people become involved in generating, gathering, and processing information to satisfy the demand for knowledge. – **microelectronic** *adj* □ see MICROCOMPUTER, SILICON CHIP, MOORE'S LAW, INFORMATION TECHNOLOGY, ROBOTICS, ARTIFICIAL INTELLIGENCE, COMMUNICATIONS SATELLITE, OPTICAL COMMUNICATIONS, FIFTH GENERATION COMPUTER

microfarad *n* (symbol μF) an electrical capacitance equal to 1 millionth (10^{-6}) of a farad □ see FARAD, CAPACITANCE

microfloppy disk *n* a floppy disk between 75 and 100 millimetres in diameter used for storing programs □ see FLOPPY DISK, STRINGY FLOPPY

micrographics system *n* an optoelectronics system for photographing and storing pages of text on microfilm. The text can be recovered from the film in printed form using a reader-printer, or, if the micrographics system is linked to a computer, the text can be recovered from memory as required. Such a system in the office takes far less space than conventional filing cabinets. □ see ELECTRONIC OFFICE

micrometre *n* (symbol μm) MICRON

micron *n* also **micrometre** (symbol μ) a distance equal to 1 millionth (10^{-6}) of a metre. The micron is a useful unit for measuring the sizes of components on a silicon chip. For example, the insulating layer of silicon oxide and the conducting aluminium tracks on a silicon chip are about 1 micron thick. The gold wire which connects the silicon chip to the terminals of an integrated circuit package is about 25 microns in diameter, and the diameter of an individual memory cell on a silicon chip is about 20 microns. □ see SILICON CHIP, PHOTOLITHOGRAPHY

Micronet 800 *trademark* a service available on British Telecom's Prestel service which enables users of some microcomputers to obtain software direct from the Prestel computer system. The programs are first downloaded into the microcomputer using a modem connected to the telephone line. From there they are saved on tape or disk for future use. Micronet 800 gets its name from page 800 on the Prestel service. □ see PRESTEL, LOCAL AREA NETWORK, MODEM

microphone *n* a device for converting sound energy into electrical energy. Several different types of microphone are available depending on the principle used for the energy conversion – compare LOUDSPEAKER □ see TRANSDUCER

microphotography *n* the use of microscopes to take photographs of small objects. Microphotography is used for showing up circuit details on silicon chips. □ see SCANNING OPTICAL MICROSCOPE, SCANNING ELECTRON MICROSCOPE

microprocessor *also* **micro** *n* a complex integrated circuit on a single silicon chip which can be programmed to carry out a wide variety of functions. Its programmability enables it to be used as the 'brain' in a microcomputer, a cash register, a washing machine, a juke box, or an industrial robot. A microprocessor or microprocessor unit comprises control circuits, registers, and an arithmetic and logic unit (ALU), but, by itself, it is a useless device: for real applications it must be combined with memory devices, which hold the instructions it operates by, power supplies, and peripheral interface devices to enable it to communicate with the real world. The control circuits hold a list of machine code instructions, known as the instruction set, which the microprocessor can understand. There are many microprocessors on the market with different instruction sets depending on the make. Four popular types of microprocessor, all operating on binary words of 8 bits in length, are the 6502, the 6800, the 8080 and the Z80. Recent 16-bit computers use either the 68000 or the Z8000 microprocessors, or designs based on them. Every microprocessor contains registers which act as temporary stores of (8-bit) words relating to instructions and addresses. The arithmetic and logic unit performs all calculations on binary numbers and makes decisions using logic gates based on instructions fed to it. All the operations performed by a microprocessor are triggered by control signals which are themselves triggered by a stream of pulses generated by a crystal clock. □ see SILICON CHIP, INSTRUCTION SET, REGISTER, CRYSTAL CLOCK, ARITHMETIC AND LOGIC UNIT, MICROCOMPUTER

microsecond *n* (symbol μs) a time interval equal to 1 millionth (10^{-6}) of a second. The time taken for a radio wave to travel a distance of 300 metres through air is about 1 microsecond.

microswitch *n* a small mechanically-operated switch usu needing only a small force to operate it. A microswitch is often incorporated into machinery (eg robotic devices) so that some moving part comes into contact with a button or lever on the switch. Although small in size, microswitches can be designed to switch several amperes of mains alternating current

microwatt *n* (symbol μW) a unit of power equal to 1 millionth (10^{-6}) of a watt

microwave oven *also* **microwave** *n* an oven in which food absorbs microwaves and becomes heated from inside, therefore cooking efficiently □ see MICROWAVES, KITCHEN ELECTRONICS

microwaves *n* radio waves having wavelengths less than about 300 millimetres and used for straight line communications and for transmitting signals through waveguides □ see MICROWAVE OVEN, WAVEGUIDE

Microwriter *trademark* a portable wordprocessor having a small set of keys which are pressed in certain combinations to produce and edit written text. A small screen shows what is being written and stored in the Microwriter's memory. The Microwriter can be linked to microcomputers for further editing, and to printers to produce hard copy. □ see WORDPROCESSOR

migration *n* **1** the movement of electrons and/or holes through a semiconductor under the influence of heat or an electric field **2** the movement of data (eg from magnetic disk to magnetic tape) in a data storage system as a result of change of use – **migrate** *vb*

millimetre *n* (symbol **mm**) a distance equal to 1 thousandth (10^{-3}) of a metre. The size of a typical silicon chip is about 5 × 5 millimetres.

millisecond *n* (symbol **ms**) a time interval equal to 1 thousandth

(10^{-3}) of a second. A laser beam takes about 1 millisecond to travel 300 kilometres through air.

milliwatt *n* (symbol **mW**) a power equal to 1 thousandth (10^{-3}) of a watt <*a 250 ~ resistor*>

minicomputer *also* **mini** *n* a computer generally regarded as having a higher performance, more high-level languages, and costing more than a microcomputer. However the border between microcomputers and minicomputers is not clear, since the performance of microcomputers is rapidly outpacing that of minicomputers. □ see MICROCOMPUTER, MAINFRAME

minority carrier *n* the least abundant of the two charge carriers, electrons and holes, in a semiconductor. Holes are minority carriers in an n-type semiconductor – compare MAJORITY CARRIER □ see SEMICONDUCTOR, N-TYPE, P-TYPE, VALENCY

MIPS [*m*illion *i*nstructions *p*er *s*econd] □ see TRANSPUTER, FIFTH GENERATION COMPUTER, MICROPROCESSOR

mission control *n* any central computing and communications facility, esp one designed to monitor and control the flight of interplanetary probes through the use of Earth stations and communications satellites □ see INTERPLANETARY PROBE

mixer *n* an electronic device for combining two or more signals. Mixers are used in sound recording when the outputs of a number of microphones are mixed to give a single output channel. They are also used in televisions and radios to produce efficient amplification of the sound and video signals. □ see HETERODYNING

mnemonic code *n* a set of computer instructions in a sort of shorthand which is easier for people to understand than machine code <*LDA,35 is a ~ meaning 'load the number 35 into register A'*> □ see ASSEMBLY LANGUAGE

MNOS – see METAL-NITRIDE-OXIDE-SEMICONDUCTOR

mode *n* a way of operating a computer so that it provides special facilities. Most computers have different display modes enabling them to display text or graphics to the best effect.

modem *n* [*mo*dulator/*dem*odulator] *also* **direct coupler** a device for converting computer data in digital form into analogue signals for transmission down a telephone line. A modem allows microcomputers to exchange games or to retrieve information from databases. □ see ACOUSTIC COUPLER

modular *adj* consisting of one or more standard units <*a microcomputer and its peripherals is a ~ system*>

modulate *vb* to change the amplitude, frequency, or phase of a wave in response to changes in some property of a message to be carried by the wave – compare DEMODULATE – **modulation** *n*, **modulator** *n* □ see AMPLITUDE MODULATION, FREQUENCY MODULATION, PHASE MODULATION UHF MODULATOR

module *n* a standard unit in an electrical system <*an amplifier ~*>

modulus *n* the number of digital states a binary counter passes through before returning to the first state <*a decade counter has a ~ of 10*>

¹monitor *n* **1** computer software, usu resident in read-only memory, for looking after various 'housekeeping' duties. These duties include displaying characters and graphics, observing which keys are pressed, and accepting data from cassettes and putting it into the right places in memory. If the housekeeping functions are more advanced than this, the monitor tends to be referred to as the operating system. □ see HOUSEKEEPING, OPERATING SYSTEM **2** a TV dedicated to use as a visual display unit and not containing the circuits for receiving TV broadcast programmes. A monitor is usu able to display graphics of higher resolution than is possible with a TV receiver since the computer can control each pixel on the screen. □ see PIXEL, RGB GUNS, UHF MODULATOR

²monitor *vb* to look at, listen into, or keep track of an operation or a system <*to ~ radio signals*>

monolithic *adj* made from a single crystal <*a ~ silicon chip*> □ see PHOTOLITHOGRAPHY, INTEGRATED CIRCUIT

¹monostable *adj* having one stable state <*a ~ multivibrator*> – compare BISTABLE □ see MULTIVIBRATOR, TIMER

²monostable *n* also **monostable multivibrator, one-shot** an electronic circuit which, when triggered, produces an output signal for a predetermined period of time before reverting back to its one normal stable state. The monostable is widely used to give a time delay in equipment ranging from washing machines to cameras and egg timers to video cassette recorders. □ see ASTABLE

Moore's Law *n* the number of components that can be integrated onto a single silicon chip will double every year. This prediction was made by Gordon Moore, founder of the Intel Corporation, in the early 1960s when integrated circuits were made from a few tens of components (known as small scale integration). To the present day Moore's prediction has been found to hold good. By the late 1960s, the annual doubling effect had led to several hundred components being integrated onto a silicon chip (known as medium scale integration). During the 1970s, component counts per chip had reached several hundred thousand (known as large scale integration). In the 1980s, microprocessor and memory chips now have close to a million components on a single silicon chip, little larger than the early chips, and this is called very large scale integration. □ see VERY LARGE SCALE INTEGRATION, SILICON CHIP, COMPUTER GENERATIONS, PHOTOLITHOGRAPHY, FIFTH GENERATION COMPUTER

Morse Code *n* a way of representing letters and numbers using combinations of dots and dashes. For sending messages by radio, the code is made up of long (dash) and short (dot) transmissions of a carrier wave. Morse Code is unsuitable for use in computers since each letter uses a different number of dots and dashes as compared with the ASCII code which uses a fixed number of binary digits for each character. □ see ASCII

MOS [*m*etal *o*xide *s*emiconductor] – see METAL-OXIDE SEMICONDUCTOR FIELD-EFFECT TRANSISTOR

MOS capacitor *n* a capacitor formed on a silicon chip from two layers of aluminium separated by a layer of silicon oxide which acts as a dielectric. This method of making capacitors on a chip during the manufacture of integrated circuits provides small capacitances of a few hundred picofarads, unlike resistors which can be made with high values of resistance. The silicon oxide and aluminium are deposited on the surface of the silicon using the process of photolithography and etching – compare JUNCTION CAPACITOR □ see DIELECTRIC, PHOTOLITHOGRAPHY, METALLIZATION

MOSFET – see METAL-OXIDE SEMICONDUCTOR FIELD-EFFECT TRANSISTOR

most significant bit *n* (abbr **MSB**) the left-most binary digit in a binary word <*the* ~ *in the word 10001100 is 1*> – compare LEAST SIGNIFICANT BIT □ see BINARY NUMBER, WORD

motherboard *n* a printed circuit board that is plugged into the back of a computer and into which can be slotted other boards (daughterboards) so that the computer can operate various peripherals □ see EDGE CONNECTOR, BACKPLANE

mouse *n* a small hand-operated box connected to a computer by a trailing wire which, when moved across the surface of a table or desk, causes a cursor to move round the screen of a VDU to select options and make decisions within a program. A mouse is fitted with a simple keypad so that when the cursor has been moved to a command or to data a key is pressed to execute the command or select the data. The keypad sometimes provides access to a limited set of characters which can be used to enhance the display. A mouse works by either sensing the movement of its wheels or by reading a grid pattern on the surface over which it moves. The use of a mouse makes it particularly easy for beginners to use a sophisticated computer system, esp a business computer where the screen can display a typical office desk on which work can be moved about and dealt with without complex keyboard operations. Home computers will increasingly make use of a mouse for inputting information to the screen. □ see ICON, DIGITIZER, LIGHT PEN, JOYSTICK, COMPUTER GRAPHICS, TURTLE

moving coil meter *n* a meter comprising a coil of fine wire which rotates in a strong magnetic field and operates a pointer moving over a scale. The movement of the coil is restrained by spiral springs or a torsion wire so that the pointer deflects by an amount proportional to the current flowing through the coil. The moving coil meter is the basis of analogue multimeters. □ see ANALOGUE MULTIMETER, DIGITAL MULTIMETER

MPU [*m*icroprocessor *u*nit] – see MICROPROCESSOR

MSB – see MOST-SIGNIFICANT BIT
MSI – see MEDIUM SCALE INTEGRATION
M/S ratio *n* MARK-TO-SPACE RATIO
multimeter *n* an instrument for measuring current, voltage, and resistance and used for testing and fault-finding in the design and use of electronic circuits. Multimeters are becoming increasingly sophisticated; some of them now include both analogue and digital displays and autoranging capability so that ranges do not have to be preselected before a measurement is taken. Many recent multimeters have become 'smart' through the use of microprocessors and thus their uses have been extended. For example, by linking a smart multimeter to an oscilloscope, measurements being taken can be shown 'live', a facility which turns the multimeter into a data logger. Other multimeters include computer interfaces so that a graphic display of measurements can be displayed on a VDU if necessary. □ see SMART, ANALOGUE MULTIMETER, DIGITAL MULTIMETER, DATA LOGGER
multiplexing *n* a method of making a single communications channel carry several messages simultaneously – compare DEMULTIPLEXING – **multiplexer** *n*, **multiplex** *vb*
multiprocessing *n* the use of more than one computer to do the same job. The additional computers are used as backups in the event of failure or to provide extra computing power. □ see MULTIPROCESSOR, MULTITASKING
multiprocessor *n* a computer system having a number of separate microprocessors cooperating with each other to process information rapidly □ see FIFTH GENERATION COMPUTER, PARALLEL PROCESSING
multistatement line *n* a program line containing more than one instruction
multitasking *n* the execution of a number of tasks simultaneously by a computer. If the computer has two or more microprocessors, simultaneous operation is truly possible. A computer with a single processor can only handle different tasks in rapid succession although they appear to be handled simultaneously. Multitasking enables a computer to make computations while waiting for slower input/output data transfers to take place □ see TIMESHARING
multivibrator *n* any of three types of two-stage transistor circuit in which the output of each stage is fed back to the input of the other by coupling capacitors or resistors, causing the transistors to switch on or off very quickly. The term multivibrator refers to this rapid switching since a square wave can be analysed into a large number of sine waves with frequencies that are multiples (harmonics) of the fundamental frequency. The three types of multivibrator are the bistable multivibrator (or flip-flop), the astable multivibrator, and the monostable multivibrator. □ see BISTABLE, ASTABLE, MONOSTABLE

music synthesizer *n* – see SYNTHESIZER, ELECTRONIC MUSIC
Mylar *trademark* – used for a plastic (polyester) film which is coated
with a magnetizable material (eg ferrite) to produce magnetic tapes
and disks □ see FERRITE, MAGNETISM, MAGNETIC TAPE, MAGNETIC
DISK

N

n the symbol for the prefix nano meaning one thousand millionth of
□ see NANOFARAD

naked *adj* supplied or used without fittings or a case *<a ~
microcomputer>*

NAND gate *n* a decision-making building block in digital circuits
which produces an output of binary 1 when one or more of its inputs
are at binary 0, and an output of binary 0 when all its inputs are at
binary 1. NAND gates are gates generally used in integrated circuit
packages. □ see GATE 1

nanofarad *n* (symbol **nF**) an electrical capacitance equal to 1
thousand millionth (10^{-9}) of a farad □ see FARAD, CAPACITANCE

nanometre *n* (symbol **nm**) a distance equal to 1 thousand millionth
(10^{-9}) of a metre

nanosecond *n* (symbol **ns**) a time interval equal to 1 thousand
millionth (10^{-9}) of a second. The time taken for sunlight to travel a
distance of 1 metre is about 3 nanoseconds.

narrowcasting *n* the broadcasting of programmes by radio, TV, or
telephone (eg Prestel) to a selected group of people □ see CABLE
TV, DIRECT BROADCAST SATELLITE, PRESTEL

NC – see NUMERICAL CONTROL

n-channel *adj, of a field-effect transistor* constructed so that current
flows through an n-type semiconductor channel between the source
and drain terminals – compare P-CHANNEL □ see FIELD-EFFECT
TRANSISTOR, NMOS

needle *n* also **stylus** that part of a dot matrix print head which places
dots on paper in an impact matrix printer □ see DOT MATRIX
PRINTER

negate *vb* to change a logic signal from one voltage level to another
(ie from 0 to 1 or vice versa) *<a NOT gate ~s a binary signal>*
□ see INVERTER, NOT GATE, OVERBAR

negative *adj* **1** charged with an excess of electrons **2** *of part of a
circuit* having a low electrical potential *<the ~ terminal of a battery>*
3 *of a number* numerically smaller than zero – compare POSITIVE
□ see CONVENTIONAL CURRENT

negative feedback *n* the feeding back to the input of part of the
output signal of an amplifier so that it subtracts from the original
input signal. Designers of amplifiers use negative feedback to great
advantage for it improves an amplifier's stability, reduces any
distortion to the signal passing through it, and increases the range of
frequencies (its bandwidth) it amplifies – compare POSITIVE
FEEDBACK

negative logic *n* a convention in the design and assembly of digital
logic circuits of letting the lower of two voltage levels represent
binary 1. Negative logic is less commonly used than the convention

of positive logic – compare POSITIVE LOGIC

negative photoresist *n* – see PHOTORESIST

nested loop *n* a routine within a routine in a computer program so that the inner routine has to be carried out every time the outer routine is repeated □ see DELAY LOOP

nesting *n* the embedding of part of a program (eg routine A) within another part (eg routine B) so that the execution of B automatically executes A – **nest** *vb* □ see NESTED LOOP

network *n* **1** any interconnected set of computer terminals that provides a data communication service (eg Prestel) **2** any organized communications system under common control *<the telephone ~>* **3** a set of similar or identical components in a single package *<a resistor ~>* **4** a complex circuit with a particular function *<a filter ~>* **5** a group of radio or television stations able to broadcast the same programmes *<the BBC ~>* □ see LOCAL AREA NETWORK

network *vb* to make programmes available to a radio or television network *< ~ed programmes>*

networking *n* the process of operating a communications system for users in different locations through a series of exchanges or terminals

neutral *adj* **1** having no electrical charge **2** of or being a conductor having zero potential

neutron *n* a particle in the nucleus of an atom which does not have an electrical charge but contributes to the mass of the atom □ see PROTON, NUCLEUS

nibble *also* **nybble** *n* half a byte, equal to a binary word of 4 bits (eg 1011) □ see BYTE, GULP

Nicad *n* [*ni*ckel-*cad*mium] a secondary cell which does not produce any gas during recharge and which, therefore, can be fully sealed. Nicads are costly but they can be recharged many times. They are widely used in portable electronic equipment (eg oscilloscopes and other test gear), and as a back-up power supply in some computers to ensure that data is not lost from memory if the mains supply fails. □ see LITHIUM CELL

night sight *n* a device usu fitted to weapons to enable objects to be seen in the very weak light from the moon and stars. A night sight uses glass lenses to form an image on a light-sensitive screen, as in a cathode-ray tube, but the image is intensified electronically so that it can be seen in the viewfinder. Some night sights respond only to the infrared light emitted from animals and enemies.

NMOS *adj, of a metal-oxide semiconductor field-effect transistor* having a conducting channel made of n-type semiconductor □ see FIELD-EFFECT TRANSISTOR, N-TYPE

node *n* **1** a point in a circuit where several conductors meet **2** a point in a communications system where information is interchanged – **nodal** *adj*

noise *n* useless and often random electrical interference on a communications link which makes a message difficult or impossible to understand □ see INTERFERENCE

noncontact *adj* having no physical contact <*a reed switch is a ~ transducer*> □ see OPTOCOUPLER, HALL-EFFECT SWITCH

noninverting amplifier *n* a voltage amplifier, usu based on an operational amplifier, which produces an output voltage directly proportional to the input voltage and of the same sign, ie a positive input voltage gives a positive output voltage – compare INVERTING AMPLIFIER □ see OPERATIONAL AMPLIFIER

nonohmic conductor *n* a conductor of electricity (eg a forward-biased pn junction in a diode) which does not obey Ohm's Law – compare OHMIC CONDUCTOR

nonvolatile *adj, of a computer memory* continuing to hold stored data whether power to it is switched on or not. Magnetic disk and tape, and bubble memory are examples of nonvolatile memory – compare VOLATILE □ see RANDOM-ACCESS MEMORY, READ-ONLY MEMORY

NOR gate *n* a decision-making building block in digital circuits which produces an output of binary 1 when all its inputs are at binary 0, and an output of binary 0 when one or more of its inputs are at binary 1. NOR gates are generally used in integrated circuit packages b □ see GATE 1

notch filter *n* a circuit designed to reject signals of a particular frequency, esp unwanted signals such as mains hum □ see FILTER, MAINS HUM

NOT gate *n* a decision-making building block in digital circuits which produces an output of binary 1 when its single input has a value of binary 0, and vice versa. NOT gates are generally used in integrated circuit packages. □ see GATE 1, INVERTER

npn transistor *n* a semiconductor device made from both n-type and p-type semiconductor and used for switching and amplifying electrical signals. The npn transistor has three terminals; the emitter and collector terminals are connected to regions of n-type semiconductor between which is a p-type region connected to the base terminal. The npn transistor is capable of amplifying a small current flowing into its base terminal. The amplified current flows between the collector and emitter terminals – compare PNP TRANSISTOR □ see BIPOLAR TRANSISTOR, FIELD-EFFECT TRANSISTOR

n-type *adj, of a semiconductor* conducting electricity by the movement of negatively charged electrons – compare P-TYPE □ see SEMICONDUCTOR, IMPURITY, VALENCY, NMOS

nuclear *adj* relating to the nucleus of an atom, atomic power, or the atomic bomb □ see NUCLEUS

nucleus *n* the central and relatively small part of an atom which is made up of protons and neutrons. The positive charge carried by the protons gives the nucleus an overall positive charge which is balanced by the negative charge on the cloud of electrons surrounding the nucleus. Atoms which lose or gain electrons are said to be ionized and have the power to attract or repel free electrons or other ionized atoms. Even neutral atoms which have no overall positive or negative charge combine together by sharing electrons. A study of the bonding and interaction between atoms helps in understanding the properties and uses of materials (eg semiconductors) which go to make up electronic components. □ see ELECTRON, SEMICONDUCTOR, VALENCY, ENERGY GAP

null *n* a zero or minimum value of an electrical signal <*a ~ in the reading of a radio direction finder*> □ see NULL-OFFSET

null-offset *n* a facility offered by some integrated circuit operational amplifiers for setting their output signal to zero to compensate for differences in the input characteristics of the amplifier. The null-offset is useful in calibrating instruments based on operational amplifiers.

number cruncher *n* any digital electronic device, esp one used for scientific analysis, which carries out many arithmetical calculations rapidly

number system *n* a set of numbers for carrying out arithmetic and logic operations. The binary, decimal, and hexadecimal number systems are used widely in computers. □ see BASE, BINARY, DECIMAL, HEXADECIMAL

numerical control *n* (abbr **NC**) the automatic control of machines (eg lathes) by means of numerical instructions in binary code. The machining operation is broken down into a series of small steps using a computer. The resulting program of instructions is then fed to servomechanisms, such as stepping motors, which control the machining operations < ~ *machine tool*>. □ see STEPPING MOTOR, COMPUTER NUMERICALLY CONTROLLED MACHINE, COMPUTER-AIDED MANUFACTURING

numeric keypad *n* a set of numbered keys which are grouped together in a separate part of a keyboard to make it easier to enter numbers into a data processing system □ see QWERTY KEYBOARD

nybble *n* NIBBLE

O

object code *n* the instructions into which a program written in source code is translated by a compiler. In many cases the object code is machine code. □ see SOURCE CODE, COMPILER, MACHINE CODE

object language *n* a language, often machine code, in which object code is expressed □ see ASSEMBLY LANGUAGE, MACHINE CODE

OCR – see OPTICAL CHARACTER RECOGNITION

octal *adj* **1** being or belonging to a number system having a number base of 8 and using the digits 0 to 7 **2** comprising a group of 8 devices *<an ~ latch>* □ see HEXADECIMAL

octave *n* a range of frequencies over which the frequency doubles *<one ~ of sound>*

off-line *adj* not controlled directly by a computer *<an ~ printer >* – compare ON-LINE – **off-line** *adv*

ohm *n* the unit of electrical resistance. A material, usu in the form of a wire, has an electrical resistance of 1 ohm if a potential difference of 1 volt across its ends causes a current of 1 ampere to flow through it. □ see OHM'S LAW, RESISTOR

ohmic conductor *n* a conductor of electricity (eg copper at constant temperature) which obeys Ohm's Law – compare NONOHMIC CONDUCTOR

ohmmeter *n* an instrument for measuring the electrical resistance of something □ see MULTIMETER, RESISTOR, OHM'S LAW

Ohm's Law *n* The potential difference across the ends of a metallic conductor is proportional to the current flowing through it if the physical state of the conductor (eg its temperature and shape) remain unchanged. Ohm's Law provides the relationship between the resistance of a material, the potential difference across its ends, and the current flowing through it, ie resistance = voltage ÷ current or R=V/I. This relationship is used by circuit designers to estimate the operating characteristics of circuits and is the most used law in electronics. □ see RESISTANCE, OHM

on-chip *adj* being on the same silicon chip as other elements *<an ~ oscillator>*

one's complement arithmetic *n* a way of subtracting one binary number from another by finding the complement of the number by changing all its 1s to 0s and vice versa and then adding this number to the number from which it was to be subtracted. If there is a carry at the left-most bit this is added to the result of the addition as follows. Suppose the subtraction is (01010111) − (00100010); using one's complement arithmetic this becomes (01010111) + (11011101) = 00110100 + 1 = (00110101)

one-shot *n* MONOSTABLE

on-line *adj* controlled directly by a computer *<an ~ printer>* –

compare OFF-LINE – **on-line** *adv*

on/off switch *n* any switch which opens and closes a circuit □ see KEY, REED SWITCH

op amp *n* OPERATIONAL AMPLIFIER

op code *n* [*op*eration *code*] the part of an instruction in machine code which tells the computer what operation is to be performed <*'jmp' is an ~ for 'jump'*> □ see OPERAND

open *adj, of a switch* having two or more contacts which are not touching so that current cannot flow in a circuit – compare CLOSED □ see OPEN CIRCUIT

open circuit *n* a circuit with an accidental or deliberate break so that current cannot flow through it – compare CLOSED CIRCUIT – **open-circuit** *vb*, **open-circuited** *adj*

open-ended *adj* not being restricted to a fixed set of data or equipment <*an ~ office system*> – **open-ended** *adv*

open-loop *adj, of a control system* having no feedback for comparison and corrective adjustment and so giving a preset output – compare CLOSED-LOOP □ see NEGATIVE FEEDBACK

open-loop gain *n* the voltage gain (amplification) of an amplifier when no signal is fed back from its output to its input. The open-loop gain of an operational amplifier is very high, being between 100 thousand and 1 million – compare CLOSED-LOOP GAIN □ see OPERATIONAL AMPLIFIER, GAIN

operand *n* the part of an instruction in machine code which tells the computer where to find the data to be worked on □ see OP CODE, MACHINE CODE, INDIRECT ADDRESSING

operating system *n* (abbr **O/S**) a special set of programs kept permanently inside a microcomputer's memory, often ROM, while the computer is running which control various housekeeping tasks (eg putting line numbers in the correct sequence when a program is entered through the keyboard). Generally, users are totally unaware that the operating system is working, ie it is transparent in operation. □ see TRANSPARENT, DISK OPERATING SYSTEM

operation *n* something (to be) accomplished; *esp* a single step to be performed by a computer in the execution of a program <*an arithmetic ~ in a microprocessor*>

operational amplifier *also* **op amp** *n* a very high gain amplifier which produces an output voltage proportional to the difference of voltage between its two inputs. Operational amplifiers were the first integrated circuits to be manufactured in large numbers in the 1960s, largely for use in analogue computers where they helped to perform mathematical operations (hence their name). Nowadays, digital computers perform most computing tasks but the operational amplifier is still in great demand by designers of audio amplifiers, instrumentation systems, voltage regulators, and control systems. □ see ANALOGUE COMPUTER, NEGATIVE FEEDBACK

operator *n* a symbol for a mathematical or logical operation <*the ~*

121

means less than>
optical *adj* of or using light *<an ~ fibre>* ☐ see OPTICAL
CHARACTER RECOGNITION, OPTICAL COMMUNICATIONS
optical astronomy *n* the use of telescopes to study planets of the
solar system, stars, galaxies, and other celestial objects from the light
they emit. Observations from earth-based observatories are limited
by absorption and distortion of light in the atmosphere, although the
use of electronic cameras and computers can vastly improve the
images obtained. Earth-orbiting observatories provide much better
'seeing' and, coupled with the latest electronic sensing, control, and
communications systems, provide more accurate observations over a
wider electromagnetic spectrum than studies from the Earth's
surface. ☐ see ELECTROMAGNETIC SPECTRUM, SPACE TELESCOPE,
RADIO ASTRONOMY
optical character recognition *n* (abbr **OCR**) a method of inputting
handwritten or printed information into a computer system by using
photodiodes to read the variation in the light reflected from the
paper ☐ see BAR CODING, LIGHT PEN
optical communications *n* also **fibre-optics communications** the use
of long thin glass fibres for sending messages using laser light.
Optical communications is fast becoming an attractive alternative to
communication by wire for several reasons: strong electric and
magnetic fields generated by lightning and electrical machinery do
not interfere with the message carried on the light beam; there is no
interference (crosstalk) between signals in neighbouring fibres;
broken fibres are not a fire hazard since the escaping light is
harmless; glass fibres are cheaper and lighter than copper wires; and,
most importantly, by using laser light, switched on and off very
rapidly in response to the messages being transmitted, a single glass
fibre can carry considerably more information than a copper wire.
The main problem at the moment is the production of glass of
sufficiently high purity to reduce the number of boosters required to
regenerate the weakening signals over long distances, but the first
submarine optical cable has been laid between Britain and Holland.
Note that conventional microelectronic circuits and devices are
needed to produce and process the information sent along optical
fibres. ☐ see LASER, OPTICAL FIBRE, ELECTROMAGNETIC PULSE
optical disk *n* – see OPTICAL MEMORY, VIDEO DISK
optical fibre *n* a thin glass or plastic thread through which light can
travel without escaping from its sides. Coloured light spilling out of
the ends of a bundle of plastic optical fibres is often used in
decorative lamps. Doctors use an optical fibre to illuminate and
examine internal organs of the body without causing too much
discomfort. But without doubt optical fibres are destined to have
their most far-reaching effects in communications systems. ☐ see
OPTICAL COMMUNICATIONS, CABLE TV
optical memory *also* **optical storage** *n* a method of storing computer

data in digital form which is read by optical means. Optical memory has not yet been developed for computer systems, but it is possible to use the 330-millimetre diameter optical disk, developed for storing TV programmes, to store over 50 thousand high quality images for easy access by a computer. The data, stored as millions of microscopic pits in the surface of the disk, is read using a laser beam. Each picture can be accessed directly and accurately using parallel access (not serial access as with a video casette recorder). The most useful use of optical memory would be in a pictorial database system, perhaps as a part of a training program, on any subject (eg natural history) which depends on visual information. There is interest, too, in developing holographic memory systems. □ see DIGITAL RECORDING, VIDEO DISK, HOLOGRAPHIC MEMORY, EXPERT SYSTEM

optical storage *n* OPTICAL MEMORY

optical wand *n* LIGHT PEN

optical waveguide *n* an optical fibre for transmitting signals using light □ see WAVEGUIDE, OPTICAL COMMUNICATIONS

optocoupler *also* **optoisolator** *n* a device for passing signals from one part of a circuit to another using a light beam. This technique ensures complete electrical isolation of the two parts of the circuit. Optocouplers are available in integrated circuit packages and are often used to couple signals from delicate low-power devices (eg microcomputers) to high-power high-voltage circuits (eg mains equipment). The signals are coupled using an infrared beam generated by a light-emitting diode and detected by a phototransistor. □ see OPTOELECTRONICS

optoelectronics *n* a branch of electronics dealing with the interaction between light and electricity. Optoelectronics has made it possible to design digital watches and clocks, electronic scoreboards, optical communications systems, and TV remote control systems – **optoelectronic** *adj* □ see LIGHT-EMITTING DIODE, LIQUID CRYSTAL DISPLAY, OPTICAL COMMUNICATIONS

optoisolator *n* OPTOCOUPLER

optoswitch *n* an (optoelectronic) device which uses a light beam to detect the movement of an object. There are two main types of optoswitch. A reflective optoswitch has a built-in light-emitting diode which sends out a beam of infrared light. If this light is reflected off a marker on a moving object it is detected by a photodiode and converted into an electrical pulse. A slotted optoswitch operates in the same way but the object moves between the light-emitting diode and photodiode so cutting off the beam. Optoswitches are generally used for measuring the rate of rotation of objects, as in anemometers and tachometers, or for monitoring the amount of rotation of a wheel, as in robotic devices. □ see OPTOELECTRONICS, OPTOCOUPLER, LIGHT-EMITTING DIODE, PHOTODIODE

Oracle *trademark* – see TELETEXT

order *vb* to arrange a set of data into a particular pattern according to some rule *<to ~ names alphabetically>* □ see SORT

OR gate *n* a decision-making building block in digital circuits which produces an output of binary 1 when one or more of its inputs has a value of binary 1, and an output of binary 0 when all its inputs have a value of binary 0. OR gates are generally used in integrated circuit packages. □ see GATE 1

O/S – see OPERATING SYSTEM

oscillation *n* a flow of electricity, or a property of a wave, which regularly changes direction *<the ~ of the mains alternating current>* – – **oscillate** *vb* □ see OSCILLATOR

oscillator *n* a circuit or device for producing a periodic waveform of specific properties (sinusoidal, triangular, square, etc). Oscillators are widely used building blocks in analogue and digital systems, esp in electronic musical instruments, digital clocks and watches, and computer systems. □ see CRYSTAL CLOCK, WAVEFORM GENERATOR

oscillograph *n* wavy traces produced on paper by a chart recorder showing the variation of some measurement (eg temperature) with time □ see CHART RECORDER, CATHODE-RAY OSCILLOSCOPE

oscilloscope *n* also **scope** CATHODE-RAY OSCILLOSCOPE

¹output *n* **1** the act or process of delivering information **2** the terminal of a device (eg an amplifier) from which information is delivered **3** a signal which represents information *<the ~ displayed on a cathode-ray oscilloscope>* – compare INPUT

²output *vb* to send information from one device to another *<to ~ data from a computer to a printer>* – compare INPUT

output device *n* a computer peripheral (eg a printer) which receives data from the computer and provides an output – compare INPUT DEVICE

output enable *n* a signal applied to an integrated circuit device (eg a memory) which allows data to be obtained from the device

overbar *n* **1** a line across the top of an abbreviation indicating that the function of the terminal of an integrated circuit is active when it is held at logic 0. Microprocessor and other chips have a number of terminals whose functions are labelled with an overbar (eg $\overline{\text{RESET}}$ means that a logic 0 signal applied to this pin causes an internal function to be set to zero). **2** a line across the top of a symbol in Boolean algebra which indicates that the logic operation the symbol indicates is to be negated, ie changed from a logic 1 to a logic 0, or vice versa. Thus, the symbol $\bar{A}$ means 'NOT A', so if A=1, $\bar{A}$=0; if A=0, $\bar{A}$=1.

overload *vb* to make a device or system exceed its operating capacity *<to ~ a heating circuit>* – **overload** *n*

overtone *n* any musical tone produced along with a fundamental frequency and which usu gives a note musical quality *<a harmonic is an ~>* □ see HARMONIC

overwrite *vb* to update data in a storage device, thus destroying the existing data □ see WRITE, REFRESH

oxide isolation *n* a way of providing electrical insulation between 'islands' of semiconductor components (eg transistors) on a silicon chip by using a layer of electrically insulating silicon dioxide – compare JUNCTION ISOLATION □ see PHOTOLITHOGRAPHY, ETCHING

P

p the symbol for the prefix pico meaning one million millionth of □ see PICOFARAD

pacemaker *n* an electronic device for stimulating a diseased or faulty heart by giving it electric shocks at regular intervals. A pacemaker is surgically fitted into the body and carries its own power supply so that it can operate unattended for some years. □ see ELECTROCARDIOGRAPH

package *n* **1** the plastic or ceramic material which is used to cover and protect an integrated circuit <*a dual-in-line ~*> **2** a computer program or set of programs with the associated documentation that is designed for a particular application (eg word processing)

package count *n* the number of integrated circuit devices on a printed circuit board □ see PACKING DENSITY

packet *n* a unit of data comprising up to about 8000 bits of digital information sent as part of a message from one user to another □ see PACKET SWITCHING NETWORK

packet switching network *n* an efficient digital communications system in which packets of data forming a message are sent through a circuit and intercepted by users as required. A physical circuit between two users is made only for the time a user 'catches' a message circulating in the system.

packing density *n* **1** the number of electronic components (eg transistors) which can be produced per unit area on a silicon chip **2** the amount of data stored per unit length on a magnetic disk

pad *n* **1** a small area near the edge of a silicon chip to which a fine (gold) wire is welded connecting the chip to the outside world □ see SPIDER, PIN **2** – see DIGITIZER 1

paddle *n* a knoblike device connected to a microcomputer which is twisted to move objects (eg balls) round the screen in games programs □ see JOYSTICK

¹page *vb* **1** to obtain information on a VDU from a database <*to ~ the Oracle*> **2** to ask for a person by name over a public address system **3** to contact people by short-wave radio transmissions <*to ~ a doctor in a hospital*>

²page *n* **1** a group of consecutive memory locations in a computer. On most microcomputers a page has 256 locations but some computers have 512 or 1024 locations **2** a complete screen of information presented on a VDU <*teletext provides ~s of information from a database*> □ see VIDEOTEXT

paint *vb* to shade with colour some area of a graphics image on a VDU □ see PALETTE, COMPUTER GRAPHICS

PAL *n* [phase *al*ternating system] a colour television broadcasting system which uses 625 lines and 50 hertz field frequency

palette *n* the range of colours a computer is able to produce on a

VDU □ see PAINT

PAM [*p*ulse *a*mplitude *m*odulation] – see PULSE CODE MODULATION

panel *n* **1** a part of the enclosure of an electronic device <*the side ~ of a radio receiver*> **2** a group of control switches, indicator lights, etc on an electronic device □ see CONSOLE

panel meter *n* a meter mounted on the front of electronic equipment showing the value of some quantity <*a power output ~*>

paper feed *n* the mechanical and electrical components for moving paper through a printer □ see PRINTER

paperless office *n* an electronic office in which the preparation, communication, and filing of messages is done electronically using wordprocessors, database systems, etc □ see ELECTRONIC OFFICE, ELECTRONIC MAIL

Papert, Seymour. Inventor of the high-level programming language Logo and author of the book 'Mindstorms – Children, Computers and Powerful Ideas'. □ see LOGO, WALTER, TURTLE

paper tape *n* one of the oldest and slowest but cheapest mediums for storing information in a computer system, consisting of strips of paper wound on a wheel which are punched with holes. When the paper is fed into the computer, the holes generate a pattern of electrical pulses representing the information store. Paper tape is not common now but its most recent use was in teleprinters.

parallel *adj* **1** of events or processes occurring simultaneously <*~ processing in a supercomputer*> – compare SERIAL **2** of or being components connected side-by-side <*a ~ circuit*> **3** of or being data transmission in which several binary digits are transmitted simultaneously

parallel circuit *n* a circuit in which components (eg capacitors) are connected side-by-side so that current is shared between each component – compare SERIES CIRCUIT

parallel input-output *n* a method of transferring data in groups of binary digits, usu in byte-sized words, between a computer and a peripheral device (eg a printer). Parallel input-output transfers data faster than serial input-output – compare SERIAL INPUT-OUTPUT

parallel interface *n* a circuit connecting a computer with a printer or other peripheral that enables parallel data transfer – compare SERIAL INTERFACE □ see CENTRONICS INTERFACE

parallel processing *n* the processing by a computer of a number of items of data simultaneously – compare SEQUENTIAL PROCESSING □ see VON NEUMANN MACHINE, FIFTH GENERATION COMPUTER

parallel processor *n* ARRAY PROCESSOR

parallel signals *n* electrical signals (eg the 8 bits on a microcomputer data bus) which travel simultaneously on separate tracks lying side-by-side – compare SERIAL SIGNALS □ see CENTRONICS INTERFACE

parity *n* a way of detecting errors in the transmission of digital data □ see PARITY CHECK

parity bit *n* – see PARITY CHECK
parity check *n* a simple method of checking whether bits in a binary word have been accidentally changed when travelling from one part of a computer to another or from one computer to another. For example, in the ASCII code the leftmost bit is a parity bit which is set to 1 if there is an odd number of 1s in the remaining 7 bits thus making 'even parity'. 'Odd parity' can also be used where the parity bit is set to make the total number of 1s odd. The byte of binary data is then checked at the receiving end against the parity bit. □ see ASCII
party line *n* a telephone line which is shared by a number of subscribers
Pascal *n* a high-level computer language named after Louis Pascal, a 17th- century mathematician, and developed during the 1960s. Pascal can deal with a wide range of problems because it is a highly structured language set out logically in blocks
pass band *n* a band of frequencies used to convey information in a communications system *<the ~ for speech in the telephone system is 4kHz>* □ see BANDWIDTH
passive *adj* of or being an electronic device or component (eg a capacitor or a resistor) which is not able to change its operating characteristics – compare ACTIVE
password *n* a set of characters, not necessarily a meaningful word, which a person must use before gaining access to a system or to information a □ see PIN
patch *vb* 1 to make a hasty correction to a computer program to cure a bug 2 to wire components together in a temporary way on a breadboard to evaluate the operation of a circuit
patchboard *n* a board containing components which are joined together using connectors on the end of wire links. An analogue computer is often wired up using a patchboard. □ see BREADBOARD, ANALOGUE COMPUTER
path *n* 1 a connecting route for signals in a circuit 2 a route between two points in a communications channel 3 a sequence of instructions in a computer program
pattern *n* a geometrical arrangement of elements *<the bit ~ of a binary word>* □ see MATRIX
pattern recognition *n* the interpretation of shapes and patterns by a computer □ see ARTIFICIAL INTELLIGENCE, FIFTH GENERATION COMPUTER
pay TV *n* TV programmes selected and paid for in the home □ see CABLE TV, DIRECT BROADCAST SATELLITE
PC – see PERSONAL COMPUTER
PCB – see PRINTED CIRCUIT BOARD
p-channel *adj, of a field-effect transistor* constructed so that current flows through a p-type semiconducting channel between the source and drain terminals – compare N-CHANNEL □ see FIELD-EFFECT

TRANSISTOR, PMOS
PCM – see PULSE CODE MODULATION
peak inverse voltage *n* (abbr **PIV**) also **maximum reverse voltage**
the maximum voltage that a diode can stand in the reverse bias
direction before it is damaged. A knowledge of peak inverse voltage
is important in the design of power supplies operating from a mains
transformer. □ see RECTIFIER DIODE, WORKING VOLTAGE
peak-to-peak voltage *also* **p/p voltage** *n* the voltage difference
between the maximum and minimum values (usu the positive and
negative values) of an alternating signal. For alternating current, the
p/p voltage is twice the peak voltage.
peak voltage *n* the maximum value reached by an alternating
current or voltage. For a 240 volt (its root mean square value) AC
mains supply, the peak voltage is $1.4 \times 240 = 340$ volts.
peek *vb* to tell a computer in the programming language Basic to
take a look at a piece of information in a memory location. The
command is followed by the address (eg A) of the location and so
the instruction is 'peek A' – compare POKE
pentavalent *adj, of an atom* having a combining power of five –
compare TRIVALENT □ see VALENCY, DOPE
pentawatt package *n* a type of encapsulation for two high-power
Darlington transistors which has five leads coming from the same
side and a metal tab for mounting the device on a heat sink
performance *n* the speed and reliability of a device <*a high ~
printer*> □ see EFFICIENCY, SPEED, RELIABILITY
period *n* the time taken for a wave motion (eg radio) to make one
complete oscillation. The period of the 50 hertz mains current is 20
milliseconds. □ see FREQUENCY
peripheral *n* also **add on, accessory** a device plugged into a
computer to expand its facilities. Peripherals include tape recorders
and disk drives for increased program storage, modems for
communicating with other computers via the telephone, and control
and data capture devices – **peripheral** *adj* □ see MODEM, INPUT/
OUTPUT PORT
peripheral interface adapter *n* (abbr **PIA**) a special integrated
circuit enabling specific microprocessors to operate external
equipment. A peripheral interface adapter usu provides two 8-bit
bidirectional lines which can be programmed by the microprocessor
for sending data to and reading from an external device. □ see
INPUT/OUTPUT PORT
permanent *adj* **1** *of a magnet* keeping its magnetism after being
magnetized **2** *of a computer memory* containing data which stays in
place when the computer is switched off <*ROM is a ~ memory*> –
see NONVOLATILE – **permanently** *adv*
personal computer *n* (abbr **PC**) a microcomputer mainly intended
for office and business use rather than for home use. Unlike a home
computer, a personal computer generally has a built-in VDU and

comes with software packages for wordprocessing functions, accessing and updating data files, and financial analysis. □ see MICROCOMPUTER

personal identity number *n* PIN

personal vital-signs communicator *n* a two-way radio communications device worn on the wrist by soldiers on the battlefield to help doctors find out to what extent the soldier might be injured. The device is under development and will contain two electrodes that will, on a signal from a battlefield station, give the soldier a mild electrical shock. The shock acts as a silent alarm. If unhurt, the soldier presses a button signalling the fact. If the soldier is injured and cannot press the button, automatic sensors go to work, monitoring heart rate, respiration rate, and the presence of muscular movement. The information goes by radio to the field hospital or command post letting doctors know the approximate extent of the injuries and the location of the soldier. □ see BIOLECTRONICS, WRISTWATCH COMPUTER, WRISTWATCH TV

PGD display *n* PLANAR GAS DISCHARGE DISPLAY

phase *n* a measure (in angular degrees or radians) of how much two waves are out of step with each other □ see PHASE MODULATION

phase-locked loop *n* (abbr **PLL**) an integrated circuit device used as an oscillator which locks onto a signal using a feedback loop. Phase-locked loops are used to maintain stable operation of FM receivers and other communications and control equipment.

phase modulation *n* a way of transmitting information that involves varying (modulating) the phase of a carrier wave (eg a radio wave) in accordance with the message being sent. In particular, when transmitting binary information the phase of the carrier wave is continually inverted to indicate the binary digits 0 and 1. □ see PHASE, AMPLITUDE MODULATION, FREQUENCY MODULATION

phase shift *n* the difference, measured in degrees or radians, between the maximum voltage and current in an alternating current circuit. Phase shift is generally caused by the effects of capacitance and/or inductance in a circuit. It can be a nuisance if it causes instability or power loss but it is made use of in filters and oscillators based on operational amplifiers.

phoneme *n* any of the smallest distinctive speech sounds of a language (in English, comprising about 40 sounds) which are put together and enhanced with pitch, volume, and stress by the human vocal tract. Systems which synthesize speech electronically model the way the human voice generates phonemes. □ see ALLOPHONE, SPEECH SYNTHESIS, FIFTH GENERATION COMPUTER

phosphor *n* the substance on the inside of a TV, VDU, or oscilloscope tube which emits light when an electron beam sweeps across it. Phosphors are chosen for their colour when excited by the electrons and the time for which they continue to emit light after the electron beam has passed on. A phosphor which emits white light for

a very short interval would be used for black and white TV tubes, while one which emits bluish-green light for a few seconds might be used for showing the trace of a heart beat on a cathode-ray tube.
□ see CATHODE-RAY TUBE, PHOSPHORESCENCE

phosphorescence *n* light emitted from some phosphors after they have been excited by an electron beam. This light is called afterglow and can be seen on a TV screen after it has been switched off in a darkened room – compare FLUORESCENCE – **phosphorescent** *adj*
□ see PHOSPHOR

photocell *n* PHOTOELECTRIC CELL

photodiode *n* a light-sensitive diode which responds rapidly to the effects of light changes. Photodiodes have two terminals, a cathode and an anode, and the current which flows between these varies uniformly with the strength of the light passing into it through a small glass window or lens at its end. Photodiodes are used for fast counting (eg of marks on a bar code), for receiving light in optical communications systems, and in cameras as sensors in light meters – compare LIGHT-EMITTING DIODE □ see LASER DIODE, LIGHT-DEPENDENT RESISTOR, PHOTOTRANSISTOR

photoelectric cell *n* a device whose electrical properties are changed by light. Two commonly-used photoelectric cells are the solar cell and the light-dependent resistor. □ see LIGHT-DEPENDENT RESISTOR, SOLAR CELL

photoelectron *n* an electron released from the surface of a metal by the action of light. Photoelectrons are made use of in image intensifiers. □ see PHOTOMULTIPLIER, NIGHT SIGHT

photolithography *n* the process of using photographic techniques and chemicals to etch a minutely detailed pattern on the surface of a silicon chip. Photolithography is applied many times in the process of making an integrated circuit. Each stage in the process involves the use of photographically-prepared plates called masks or photomasks. Each mask holds a particular pattern identifying individual transistors, conducting pathways, etc. A mask is placed over a thin layer of photoresist covering the surface of the silicon. Ultraviolet light shone on the mask passes through the clear areas but is stopped by the opaque areas. According to the type of photoresist used, either the exposed or the unexposed photoresist can be dissolved away using chemicals to leave a pattern of lines and holes. This pattern enables transistors to be formed in the silicon and aluminium interconnections to be made between them. □ see PHOTOMASK, PHOTORESIST, WAFER, SILICON CHIP, ETCHING, PRINTED CIRCUIT BOARD

photomask *n* a transparent glass plate used in the manufacture of integrated circuits on a silicon chip, on which a very precise pattern of microscopically small opaque spots has been produced by photographic reduction of a much larger pattern. The pattern represents the circuit layout and defines the areas on the chip into

which impurities are to be diffused, metal is to be cut away, etc.
□ see PHOTOLITHOGRAPHY, WAFER, SILICON CHIP, METALLIZATION

photomultiplier *n* a device which increases the brightness of an image (eg in a night sight on a weapon) by increasing the number of electrons produced by the action of light on a metal in a vacuum
□ see PHOTOELECTRON, NIGHT SIGHT

photon *n* the smallest 'packet' or quantum of light energy. Light sensors such as photodiodes absorb photons and produce electrons which enable the light to be detected. □ see PHOTODIODE, LIGHT-EMITTING DIODE, QUANTUM, PLANCK'S CONSTANT

photoresist *n* also **resist** a light-sensitive material which is spread over the surface of a silicon wafer from which silicon chips are being made, and whose solubility in various chemicals is altered by exposure to light. Positive photoresist becomes more soluble when exposed to light but resists the action of an etching material on those areas which have not been exposed. Negative photoresist hardens on exposure to light thus acting as a barrier to the etching material in exposed areas. Positive resist is also used widely in the making of printed circuit boards. □ see ETCHING, PHOTOLITHOGRAPHY, PHOTOMASK

photosensitive *adj* responding to light <*a solar cell is a ~ device*> – **photosensitivity** *n* □ see PHOTOTRANSISTOR

phototransistor *n* a transistor which responds to light and produces an amplified output signal. Phototransistors have a rapid response to light and are used as receivers in optical communications systems
□ see PHOTODIODE

photovoltaic *adj* of or being a device (eg a solar cell) which produces electrical energy when it absorbs light □ see SOLAR CELL

PIA – see PERIPHERAL INTERFACE ADAPTER

picel *n* [*pic*ture *cel*l] PIXEL

pickup *n* **1** the stylus and supporting device on a record player which converts the movement of the stylus into electrical signals **2** unwanted electrical interference on a radio or other communications system

picofarad *n* (symbol **pF**) an electrical capacitance equal to 1 million millionth (10^{-12}) of a farad □ see CAPACITANCE, FARAD

picture frequency *n* the number of times per second that a set of lines forming a complete TV picture is drawn on a TV screen. For the 625 line system, the picture frequency is 25 hertz (25Hz). The picture frequency is exactly half the field frequency which is the number of times per second that the sequence of scanning operations occurs in building up a TV picture. □ see VIDEO, TELEVISION

piezoelectricity *n* the electricity that certain crystals (eg quartz) produce when they are squeezed. The effect is put to good use in some types of gas lighters, in most hi fi pickups, and in very stable electronic oscillators – **piezoelectric** *adj* □ see CRYSTAL CLOCK

pin *n* any of usu several projecting pieces of metal on a device that enables the device to be plugged or soldered into a circuit <*a transistor* ~> □ see TERMINAL 2

PIN *n* [*personal identity number*] a codeword used by a person to gain access to the facilities offered by a computer □ see ELECTRONIC FUNDS TRANSFER, SMART CARD, TELEBANKING

pinout diagram *n* a diagram showing how the internal building blocks (eg logic gates) of an integrated circuit device are connected to its pins

PIO – see PROGRAMMABLE INPUT/OUTPUT

piracy *n* the illegal copying of computer software or sound and video programmes which are protected by copyright rules □ see DONGLE

pitch *n* the highness or lowness of a sound <*a high frequency sound has a high* ~> □ see AMPLITUDE, FREQUENCY

PIV – see PEAK INVERSE VOLTAGE

pixel *also* **picel** *n* the smallest bit of light-emitting phosphor making up the image on a VDU. There are about 230 000 pixels on the screen of a monochrome display, each of which can be controlled by some microcomputers to produce high resolution pictures. □ see MONITOR, RGB GUNS

PLA – see PROGRAMMABLE LOGIC ARRAY

planar *adj* of or being a semiconductor device having one or more active components formed on a substrate of silicon <*a* ~ *transistor*>

planar diffusion *n* GASEOUS DIFFUSION

planar gas discharge display *also* **PGD display** *n* a disp·y device which produces highly visible yellow-orange light on a high-contrast black background. Either dot matrix alphanumeric characters or graphic patterns can be displayed on a panel which is easily interfaced to portable microcomputers, point of sale terminals, and banking terminals. □ see DOT MATRIX DISPLAY, SEVEN SEGMENT DISPLAY, LIQUID CRYSTAL DISPLAY

Planck's constant *n* the factor by which the frequency of electromagnetic radiation must be multiplied in order to obtain the energy of a quantum of the radiation. Planck's constant is equal to 6.62×10^{-34}watt sec^2 □ see QUANTUM, QUANTUM THEORY

platen *n* the roller in a printer or typewriter round which the paper is rolled

PLL – see PHASE-LOCKED LOOP

plot *vb* **1** to print graphics on paper using a printer □ see GRAPH PLOTTER **2** to 'light up' a pixel on the screen of a VDU by a command (eg 'plot') in a computer program. The pixel is usu one of 64 in an 8 by 8 matrix of pixels and the command has to be followed by the x and y co-ordinates of the position.

plotter *n* – see GRAPH PLOTTER

plug *n* a connector at the end of a lead which mates with a socket <*a mains* ~ *on a microcomputer*> – **plug** *vb* □ see DIN PLUG

PMOS *adj, of a metal-oxide semiconductor field-effect transistor* having a conducting channel made of p-type semiconductor □ see

FIELD-EFFECT TRANSISTOR, P-TYPE
pn junction *n* the boundary formed between a p-type and an n-type semiconductor. The electrical properties of the pn junction are the basis of operation of all semiconductor devices from the diode to the integrated circuit. Its basic function is to allow current to flow easily from the p-type to the n-type material but to stop current flowing through it in the opposite direction. Thus its function is the basis of the action of a diode. In a bipolar transistor, two pn junctions are used. □ see P-TYPE, N-TYPE, REVERSE BIAS, FORWARD BIAS, NPN TRANSISTOR
pnp transistor *n* a semiconductor device made from both p-type and n-type semiconductor and used for switching and amplifying electrical signals. The pnp transistor has three terminals; the emitter and collector terminals are connected to regions of p-type semiconductor between which is an n-type region connected to the base terminal. The pnp transistor is capable of amplifying a small current flowing into its base terminal. The amplified current flows between the collector and emitter terminals. The pnp transistor is less commonly used than the npn transistor – compare NPN TRANSISTOR □ see BIPOLAR TRANSISTOR, FIELD-EFFECT TRANSISTOR
pocket *adj, of an electronic device* small enough to be put in a pocket <*a ~ calculator*>
pocket computer *n* a microcomputer or a programmable calculator small enough to fit into the pocket □ see PORTABLE MICRO
point contact diode *n* an early type of semiconductor diode, still in use, in which a rectifying pn junction is formed at the point of contact between a fine wire and a small piece of semiconductor material (eg germanium) □ see RECTIFIER DIODE
pointer *n* PROGRAM POINTER
point-of-sale terminal *also* **POS terminal** *n* a terminal where goods are sold in supermarkets and garages. A point of sale terminal supplies details about the cost and type of purchases, and includes a printer to provide a customer with a receipt. Some contain a memory to record transactions, and a day's takings may be read from this memory by a central computer. □ see SMART CARD, ELECTRONIC FUNDS TRANSFER
poke *vb* to tell a computer in the programming language Basic to store a piece of information in a memory location. The command is followed by the address (eg A) of the location and the number (n) to be stored so the instruction is 'poke A, n' – compare PEEK
polarity *n* the electrical condition of an atom, molecule, or electrical component which makes it have negative and/or positive properties <*the ~ of a battery*> □ see ELECTROLYTIC CAPACITOR
polarize *vb* **1** to produce polarity in something <*to ~ a capacitor*> **2** to give light a 'one-sidedness' using a polarizing filter (eg so that its intensity can be reduced using a second polarizing filter)
polarizing filter *n* a transparent often coloured sheet of plastic

which is placed over a display (eg a seven-segment display) to cut down glare and make the display easier to read □ see ANTIGLARE SURFACE

pole *n* **1** either of the two ends of a magnet where magnetism is concentrated *<the North ~>* **2** either of the two terminals of a cell or battery *<the positive ~>* **3** a terminal on a switch (eg a relay) which makes electrical contact with one or more other terminals when the switch is operated

polling *n* the successive examination by a computer of a number of interfaces or terminals connected to it, to find out which has any data for it to use – **poll** *vb*

port *n* a place on a microcomputer to which peripherals can be connected to provide two-way communication between the computer and the outside world □ see INPUT/OUTPUT PORT, PERIPHERAL INTERFACE ADAPTER

portability *n* the ability to use software on more than one computer system. Portability is difficult to achieve for two reasons. First, different microcomputers use different microprocessors (eg 6502 and Z80), and these have different instruction sets. Thus a compiler used in one microcomputer to convert from a high level language to machine code works quite differently on another computer. Second, even microcomputers using the same microprocessor have their own unique addresses for screen memory, cursor control, etc. However, to overcome these difficulties, special programs have been developed, the best known of which is called CP/M. □ see CP/M, INSTRUCTION SET, COMPILER

portable *adj* capable of being carried about *<a ~ wordprocessor>* – **portability** *n*

portable micro *n* a small battery-operated microcomputer which has a built-in screen and a microcasette player for storing programs. Its internal memory is usu nonvolatile so that it retains its contents when its battery is switched off. Its screen is usu a liquid crystal display which holds a few lines of program instructions and limited graphics. Portable micros can be easily used on the lap and they are usu of interest to executives on the move, and to technicians and engineers on the factory floor or building site. When connected to a modem, a portable micro can communicate the results of field work to an office computer by the telephone line. Similarly, a salesman finds a portable micro useful for taking and processing orders. A range of peripherals, esp wordprocessors and printers, are available, and the development of voice input/output devices is giving more and more people powerful computing facilities at any time and place. □ see MICROCOMPUTER

positive *adj* **1** having a deficit of electrons **2** *of a part of a circuit* having a relatively high electrical potential *<the ~ terminal of a battery>* **3** *of a number* numerically greater than zero – compare NEGATIVE

positive feedback *n* the feeding back to the input of part of the output signal of an amplifier so that it adds to the original input signal. An amplifier can go into unwanted oscillation if positive feedback occurs accidentally. But positive feedback is put to good use in the design of oscillator circuits based on op amps, and in Schmitt trigger circuits – compare NEGATIVE FEEDBACK □ see OSCILLATOR, SCHMITT TRIGGER

positive logic *n* a convention in the design and assembly of digital logic circuits of letting the higher of two voltage levels represent binary 1. Positive logic is more commonly used than the convention of negative logic – compare NEGATIVE LOGIC

positive photoresist *n* – see PHOTORESIST

post *n* a pin or rodlike terminal enabling connections to be made to a circuit board □ see PIN

POS terminal *n* POINT-OF-SALE TERMINAL

potential *n* the electrical pressure at a point in a circuit which makes it possible for current to flow from that point to a point of lower potential □ see ELECTROMOTIVE FORCE

potential barrier *n* a narrow region across a semiconductor pn junction which opposes the movement across it of holes and electrons. Many semiconductor devices, esp the diode, depend on the creation of a potential barrier to control the flow of electricity in a circuit. □ see DEPLETION REGION, SPACE CHARGE, PN JUNCTION

potential difference *n* the difference in electrical pressure, measured in volts, between two points in a circuit which makes current flow between them b □ see ELECTROMOTIVE FORCE

potential divider *n* an electrical component consisting of one or more resistors connected in series through which current flows to produce one or more voltages, each of which is less than the total voltage across the resistors. The potential divider is a very common building block in circuit design and it is built into many integrated circuits to provide the correct voltages to operate devices (eg a transistor in an amplifier). □ see POTENTIOMETER, WHEATSTONE BRIDGE

potentiometer *n* an electrical component having three terminals used for providing an adjustable potential difference for control purposes *<a ~ as a volume control>* □ see POTENTIAL DIVIDER

¹power *n* **1** the number of times that a number is to be multiplied by itself □ see EXPONENT **2** (abbr **PWR**) the rate at which energy is absorbed or radiated by a device *<the ~ of a laser>* □ see WATT

²power *vb* to supply electrical power to a device *<to ~ an amplifier>*

power amplifier *n* an amplifier which increases the power of an electrical signal *<a hi fi audio ~>* □ see PUSH-PULL AMPLIFIER

power dissipation *n* the power generated as heat in a device owing to the current flowing through it. Every electronic device has a maximum rating of power dissipation and to exceed this maximum may damage the device. □ see HEAT SINK, THERMAL RUNAWAY

powerful *adj, of a computer or microprocessor* able to carry out many instructions at high speed <*a transputer is a ~ microprocessor*> – **powerfully** *adv*

power pack *n* POWER SUPPLY

power point *n* an outlet on equipment, a bench, a wall, or at any other accessible point which provides electrical power to equipment plugged into it <*a mains ~*>

power supply *also* **power pack** *n* a unit that is usu portable but is sometimes built into equipment for providing electrical power usu at a range of selectable voltages <*a DC ~*> □ see BALANCED POWER SUPPLY

power transistor *n* any transistor which is able to control currents in excess of 1 ampere. A power transistor has a built-in heat sink and one or more holes to enable it to be bolted down to an additional heat sink if required. Power transistors are used in power supplies, audio amplifiers, and computer interface equipment.

power up *vb* to switch on power to a device <*to ~ a microcomputer*>

p/p voltage *n* PEAK-TO-PEAK VOLTAGE

preamplifier *also* **preamp** *n* an amplifier which increases the power and voltage of a weak signal from a source (eg a magnetic pickup) to a level which may be amplified further. In an audio system a preamplifier usu includes equalization and tone control circuits.

precision *n* the degree of discrimination in a measurement <*the ~ of an 8-bit analogue-to-digital converter is 1 part in 256*> – compare ACCURACY – **precision** *adj* □ see RESOLUTION

predefined *adj* given a purpose by someone else <*~ function keys on a microcomputer*>

preferred values *n* a range of component values, esp the values of resistors, which ensures that circuit designers can find a component of suitable value within an acceptable tolerance. Thus there are 12 preferred values in the E12 series of resistors having values in the ratios 1, 1.2, 1.5, 1.8, 2.2, 2.7, 3.3, 3.9, 4.7, 5.6, 6.8, 8.2. Since this series is manufactured with a tolerance of ±10%, a continuous spread of values of resistors is available which is quite acceptable to circuit designers, since exact values of resistors are seldom necessary. There are 24 preferred values in the E24 series of resistors which are manufactured with a tolerance of ±5%.

prescaler *n* an integrated circuit device which divides the frequency of pulses by a certain factor so as to give a lower output frequency which can be handled by a counter or other device to which the pulses are fed. A prescaler is often used with a digital frequency meter so that a higher frequency range can be measured by the meter. □ see COUNTER, DIGITAL FREQUENCY METER

preset *n* an electronic component (eg a variable resistor) which is adjusted to a fixed value in a circuit – **preset** *vb*, **preset** *adj*

Prestel *trademark* a service provided by British Telecom in the UK enabling telephone subscribers to retrieve information from a computer database. The information is displayed on a VDU using a videotext terminal, or on a television receiver connected to the telephone system by means of a videotext decoder. Over a quarter of a million pages of information are available from a wide range of information providers. Prestel, like all viewdata systems, is menu-driven so that a range of choices is presented on-screen. The user makes a choice by entering a number on an alphanumeric keypad which transfers control to a selected sub-menu, and the procedure is repeated until the user arrives at the page of information required. Prestel offers information ranging from news and weather to business and financial matters. Goods may be ordered from home and travel bookings made and the service is continually expanding as more information providers enter the system. □ see VIDEOTEXT, MICRONET 800, TELESHOPPING

primary cell *n* an electric cell that cannot be recharged – compare SECONDARY CELL

primary coil *also* **primary winding** *n* the winding on a transformer that has the inducing current flowing through it – compare SECONDARY COIL □ see TRANSFORMER

primary colours *n* the three colours red, green, and blue, particular combinations of which are used to make all other colours. A colour television screen contains three phosphors which emit the primary colours when excited by the relevant electron beam. □ see RGB GUNS, COLOUR TELEVISION

print *vb* to produce characters on paper <~ing *a list of program instructions*> □ see PRINTER

printed circuit board *n* (abbr **PCB**) a thin board made of electrical insulating material (usu glass fibre) on which a network of thin copper tracks are formed to provide connections between components soldered to the tracks. The microprocessor, memory chips, and other components in a microcomputer are assembled on one or more printed circuit boards. □ see ETCHING

printer *n* a device for printing out paper copies of programs, data, etc under the control of a computer □ see DAISYWHEEL PRINTER, DOT MATRIX PRINTER, INK-JET PRINTER

printer buffer *n* – see BUFFER 2

printer port *n* a D-type or DIN-type socket on a microcomputer into which a printer can be plugged

print head *n* that part of a printer (eg a daisywheel) which does the printing

printout *n* program listings, graphics, etc printed out on paper by a printer

printwheel *n* – see DAISYWHEEL PRINTER

probe *n* any electronic device which is placed near to the point or area of investigation <*a space* ~> □ see INTERPLANETARY PROBE

procedure *n* **1** a subroutine in a program which instructs the computer to carry out one particular task **2** the sequence of steps for solving a problem *<an algorithm is a ~>*

process *vb* to analyse data or work through a set of instructions using a computer *<~ing data from seismic surveys>*

process control *n* the use of computers for controlling industrial processes (eg the production of steel) □ see REAL TIME CLOCK

processing speed *n* the rate at which a microprocessor executes a set of instructions □ see MIPS

processor *n* a central processor unit or a microprocessor □ see MICROPROCESSOR

program *n* a list of instructions written in a language that can be translated into something a computer can understand. A program is usu entered from a keyboard and stored in the computer's random-access memory, or it is loaded into it from an external program store (eg a magnetic tape or disk). □ see LIST, RUN, SOFTWARE

program counter *n* a special register in a microprocessor which contains the address of the next instruction to be carried out

program development *n* the process of writing, debugging, and testing a computer program for a particular application

programmable *adj* able to be controlled by means of a program *<a ~ central heating controller>* □ see READ-ONLY MEMORY

programmable calculator *n* – see ELECTRONIC CALCULATOR

programmable input/output *adj* (abbr **PIO**) of or being a microcomputer having a range of facilities for dealing with data input and output *<a ~ chip>* □ see PERIPHERAL, INPUT/OUTPUT CHIP

programmable logic array *n* (abbr **PLA**) a type of read-only memory that has been programmed to perform a number of logic functions. For example, a simple programmable logic array in an integrated circuit package has three terminals for selecting the three functions AND, OR, and NOT, and two inputs on which these functions can be performed as required. □ see UNCOMMITTED LOGIC ARRAY

programmable read-only memory *n* (abbr **PROM**) a type of read-only memory which can be programmed by the user □ see READ-ONLY MEMORY

programmer *n* a person who prepares programs for computers □ see PROGRAM

programming language *n* a language used for writing computer programs □ see LANGUAGE

program pointer *n* a pair of registers specified in a machine code program which tells the computer where it should store or fetch data. A program pointer is used in the programming method called indirect addressing. □ see INDIRECT ADDRESSING

Prolog *n* [*pro*gramming *log*ic] a high-level computer language, developed by Britain and France in the 1970s, and having some of the characteristics of human logic. It is particularly suitable for accessing databases and for educational applications, and it is well

suited to the needs of fifth generation computers. □ see FIFTH
GENERATION COMPUTER, ARTIFICIAL INTELLIGENCE

PROM – see PROGRAMMABLE READ-ONLY MEMORY, READ-ONLY
MEMORY

PROM burner *n* a device for altering the contents of a (fuse)
programmable read-only memory (PROM) by selectively burning
out, (ie fusing with a high voltage) diodes or transistor elements on
the chip to write 0-bits where required □ see FUSE PROM, READ-
ONLY MEMORY

prompt *n* a symbol (eg a question mark) appearing on a VDU
during the running of a program which requests an input from the
keyboard □ see ERROR CODE

propagate *vb* to send information through something *<radio waves
can be ~d through space>*

propagation *n* the travelling of waves through something *<the ~ of
light waves through an optical fibre>*

propagation speed *n* the time taken for a change to appear in the
output of a logic gate after a change in input signal. Propagation
speed is one way of comparing the performance of digital logic
devices (eg the propagation speed of TTL logic is about five times
that of CMOS logic). □ see TRANSISTOR-TRANSISTOR LOGIC,
COMPLEMENTARY METAL-OXIDE SEMICONDUCTOR LOGIC

protection *n* any method (eg the use of a dongle) for protecting
software from illegal copying □ see DONGLE, PIRACY

protocol *n* a set of (internationally) agreed rules for determining the
way information flows in a communications system. The rules
broadly cover three aspects of the system: the language by which the
information is conveyed, the meaning attached to the language, and
the sequence in which the information is transmitted. □ see RS-232
INTERFACE

proton *n* a particle that makes up the nucleus of a hydrogen atom,
coexists with neutrons in the nuclei of all other atoms, and has a
positive charge equal in value to the negative charge on an electron
□ see NUCLEUS, NEUTRON

prototype *n* the first working circuit or system on which later
designs are based *<a ~ of a robot>*

prototype board *n* a circuit assembly system (eg a breadboard) for
testing the operation of a circuit before producing a permanent
assembly □ see BREADBOARD, PATCHBOARD

proximity detector *n* an electronic device that can detect the
presence of a person (eg a burglar) without any physical contact
being involved. Proximity detectors make use of sound, light,
ultrasonic waves, infrared light, magnetism, or radio waves. □ see
SECURITY ELECTRONICS, REED SWITCH, HALL EFFECT

PSU [*power supply unit*] – see POWER SUPPLY

p-type *adj, of a semiconductor* conducting electricity by the
movement of positively charged holes – compare N-TYPE □ see

SEMICONDUCTOR, HOLE, IMPURITY, PMOS

pull-up resistor *n* a resistor used to raise the voltage at some point in a circuit (eg to maintain a voltage level of binary 1 in a digital logic circuit) □ see RESISTOR, GATE 1

pulse *n* a short-lived variation of voltage or current in a circuit, or of an electromagnetic wave (eg laser light), which generally has a precisely defined amplitude □ see PULSE CODE MODULATION, ELECTROMAGNETIC PULSE

pulse amplitude modulation *n* (abbr **PAM**) – see PULSE CODE MODULATION

pulse code modulation *n* (abbr **PCM**) a way of changing an analogue signal into a digital signal for transmission along telephone lines or by radio waves. The analogue signal (eg voice signals from a microphone) is sampled to determine the amplitude of the signal at very short intervals of time, a process known as pulse amplitude modulation. At least 8000 samples need to be taken per second to represent a speech pattern and 40 000 to represent music. Each of the samples is converted into a binary code (usu of 8 bits) and this code is the information transmitted. At the receiving end, the digital code is changed back into an analogue signal. Pulse code modulation is used in preference to analogue transmission since it is possible to recover the original signal more accurately at the receiving end, even though it may be immersed in a lot of background noise. □ see DIGITAL RECORDING, SAMPLE

pulse height *n* the amplitude, measured in volts, of a signal delivered by a detector (eg to an amplifier) □ see AMPLITUDE, PULSE

pulse shaping *n* the use of electronic circuits to improve the shape of a pulse (eg to 'square up' a ragged pulse shape) □ see SCHMITT TRIGGER

pulse train *n* a chain of electrical signals of precise shape which carry information or which are used to control the flow of information

punched card *n* a now largely obsolete medium for inputting and storing computer data that consists of a cardboard card in which a pattern of holes has been cut to represent instructions or information □ see PAPER TAPE

push-button *n* a mechanical switch which is pressed to enter data into an electrical system (eg a telephone) □ see KEYBOARD

push-button telephone *n* a telephone in which a number is selected by pushing buttons on a keypad rather than by rotating a dial. Modern push-button telephones are able to store numbers so selected for future use.

push-pull amplifier *n* an audio frequency amplifier (eg a hi-fi amplifier) which uses one npn and one pnp transistor coupled together to deliver an amplified signal to a loudspeaker or other power device efficiently. One transistor 'pushes' while the other

'pulls' the amplified signal through the loudspeaker. ☐ see
CROSSOVER DISTORTION

PVC *n* [*poly*vinyl *c*hloride] an electrically insulating plastic used as
electrical binding tape and to cover wires

PVC tape *n* a strip of sticky-backed PVC widely used as an electrical
insulator (eg to bind bare wires) ☐ see INSULATOR

pyrometer *n* an electronic instrument for measuring temperatures
above about 600C using, for example, a thermocouple as a sensor
☐ see THERMOCOUPLE

Q

Q-factor *n* the sharpness (or 'quality') of an electronic filter circuit in selecting a particular signal for amplification or rejection. A tuned circuit in a radio receiver has a high Q-factor if it separates stations so that there is no overlap between them. □ see TUNED CIRCUIT, FILTER, SELECTIVITY

QIL – see QUAD-IN-LINE

quad *adj* having four parts <*a ~ 2-input NAND gate*> □ see QUADRAPHONIC

quad amplifier *n* an integrated circuit containing four operational amplifiers in a single encapsulation

quad-in-line *adj* (abbr **QIL**) *of an integrated circuit package* having connecting pins coming out of each side of the body, but with the ends of every alternate pin bent in opposite directions so as to form 4 lines of pins. The quad-in-line package is not as common as the dual-in-line package – compare DUAL-IN-LINE

quadraphonic *adj* of or being a high fidelity audio system which uses four loudspeakers to create improved three-dimensional sound – **quadraphony** *n*, **quadraphonics** *n*

quantum *n* the smallest 'packet' in which radiant energy can be transmitted from place to place. The energy of a quantum is proportional to the frequency of the radiation and is given by the simple equation: quantum energy = Planck's constant × frequency □ see QUANTUM THEORY, PLANCK'S CONSTANT, ELECTRON VOLT

quantum theory *n* a theory of physics which is able to explain very successfully many of the complex ways energy and matter interact by assuming that electromagnetic radiation (eg light) comprises 'packets' of energy. Quantum theory has been very useful in understanding and predicting the behaviour of semiconductors. □ see PLANCK'S CONSTANT, QUANTUM, SEMICONDUCTOR PHYSICS

[1]quartz *n* a crystalline form of silicon oxide which has two main properties which are of use in electronics: (1) it is piezoelectric so that it can be used for the generation of precise timing signals in clocks and watches <*a ~ crystal watch*>; (2) it is transparent to ultraviolet light so that it can be used as a window covering an electrically programmable read-only memory to allow ultraviolet to erase the contents of the memory □ see CRYSTAL CLOCK

[2]quartz *adj* controlled by the oscillations of a crystal of quartz <*a ~ watch*> □ see CRYSTAL CLOCK

queue *n* a waiting line of data in a computer memory which is processed according to the 'fifo' (first in first out) principle (ie the first data or instruction entered is also the first to be removed)

quiescent state *n* the ('quiet') state of a circuit when it is not active (eg amplifying) though it is in perfect working order and connected to a power supply

qwerty keyboard *n* a computer keyboard which has its keys arranged in the same way as those of a standard typewriter. There may be small variations in the layout of the complete keyboard but the first six letters on the top row of keys should spell 'qwerty'.
□ see KEYBOARD

R

R the symbol for electrical resistance □ see RESISTANCE

radar *n* [*r*adio *d*etection *a*nd *r*anging] an electronic system which sends out high frequency radio waves to locate distant objects by their echo <*a ship's* ~>. Radar has been used by astronomers to obtain images of the surface of cloud-covered Venus. – **radar** *adj* □ see VENUS RADAR MAPPER

radiate *vb* to send out energy as light, heat, radio waves, or as any other form of electromagnetic radiation <*a transmitter aerial* ~s *radio waves*>

radiation *n* energy transmitted through the air, space, or other medium as electromagnetic waves <*laser* ~> □ see ELECTRO-MAGNETIC WAVE

¹radio *vb* to send information by radio waves

²radio *n* communication at a distance by means of electromagnetic waves having frequencies between about 15 kilohertz and 100 megahertz □ see LONG WAVES, MEDIUM WAVES, SHORT WAVES, MICROWAVES, GAMMA RAYS, ELECTROMAGNETIC WAVES

radio astronomy *n* the use of radio receiver equipment and specially designed aerials to study the behaviour of the Sun, stars, and galaxies. Radio astronomy and optical astronomy are the two main ways we receive electromagnetic waves from these distant objects through narrow windows in the Earth's atmosphere. Both Earth-based and space-based radio telescopes, using the latest advances in electronics, have greatly expanded our knowledge about the structure and origins of the universe. □ see ELECTROMAGNETIC SPECTRUM, ELECTROMAGNETIC WAVE, DISH AERIAL, MASER, OPTICAL ASTRONOMY, WINDOW 1

radio frequency *n* (abbr **RF**) a frequency in the range 15 kilohertz to 900 megahertz used for communication □ see LONG WAVES, MEDIUM WAVES, SHORT WAVES, MICROWAVES

radio frequency generator *n* an oscillator generating radio waves covering the frequency range from about 100 kilohertz to about 300 megahertz. Radio frequency generators are used in radio circuit design and testing and their output is generally a sine wave which can be frequency and amplitude modulated.

radio microphone *n* an ordinary microphone with a low-power FM (frequency modulated) transmitter built into its handle from which a short wire aerial hangs. The frequency of transmission can be adjusted by means of a screw so that an ordinary FM receiver can be tuned to pick up a transmission in the frequency range 88 megahertz to 108 megahertz. Radio microphones are widely used in TV studios to allow people to move around unencumbered by trailing wires, and to avoid the need for a microphone slung at the end of a boom.

radiosonde *n* a compact instrumentation package borne aloft by balloon to measure weather conditions in the atmosphere and used in weather forecasting. It comprises instruments for measuring temperature, pressure, relative humidity, etc, and a radio transmitter for transmitting the data to a meteorological station. □ see TELEMETRY

radix *n* BASE 1

RAM – see RANDOM-ACCESS MEMORY

ramp generator *n* a circuit which generates a waveform shaped like a sawtooth. Ramp generators are used to move a small spot produced by an electron beam at uniform speed across the screens of VDUs, TVs, and oscilloscopes. □ see RASTER

RAMTOP *also* **HIRAM** the highest address at which program information can be stored in the random-access memory of a microcomputer

random-access *adj, of a file, memory device, etc* containing data that can be found straight away wherever it is – compare SEQUENTIAL ACCESS □ see RANDOM-ACCESS MEMORY, ACCESS

random-access memory *n* (abbr **RAM**) a semiconductor device, usu consisting of one or more integrated circuits, which is used for the temporary storage of data in a computer system. All microcomputers use RAM to store data and programs entered from the keyboard or loaded into it from an external store (eg magnetic tape or disk). The data is usu stored in a RAM in the form of 8-bit words (bytes). The storage capacity of a RAM is usu measured in thousands of bytes (eg 48K). The data is held in storage locations identified by addresses and can be 'written into' or 'read from' these locations in any order at random – compare READ-ONLY MEMORY □ see DRAM, STATIC

random number *n* any number in a set that is impossible to predict. All microcomputers generate random numbers for use in computer games of chance and in the simulation of randomly fluctuating processes (eg radioactive decay of an isotope).

range *n* the difference between the lower and upper values of a quantity <*the ~ of audio frequencies handled by an amplifier*>

raster *n* the pattern of electrical signals produced when an electron beam scans the image of a picture inside a video camera. This pattern forms the video signal which is broadcast as a modulated radio frequency carrier wave. □ see VIDEO CAMERA, RASTER SCANNING

raster scanning *n* the way a beam of electrons is swept back and forth across the screen of a VDU or TV to build up a picture. The beam scans back and forth, from the top left-hand corner to the bottom right. □ see TELEVISION

rate *n* the number of events per second, minute, or hour <*bit ~*> □ see BAUD, FREQUENCY

ray *n* a narrow beam of electromagnetic radiation which travels in a straight line <*gamma ~s*> □ see LASER, X-RAYS

reactance *n* the resistance (in ohms) due to capacitance or inductance in an alternating current circuit <*capacitive* ~> □ see IMPEDANCE

reactive *adj* providing a resistance to the passage of alternating current <*a* ~ *circuit*>

read *vb* to sense and deliver information to a computer <*to* ~ *data from a floppy disk*> – compare WRITE □ see READ/WRITE HEAD, READ-ONLY MEMORY

read head *n* a magnetic device in tape and disk drives which reads data from a magnetic tape or disk – compare WRITE HEAD

reading machine *n* an electronic device for converting the printed word into synthetic speech. A reading machine is useful to blind people who do not have to master the special skills required in the use of Braille and other tactile devices. Reading machines consist of three parts: a scanner unit which forms an optical image of the printed page line by line; an electronic control unit based on a microprocessor, memory, and a tape drive unit for the program of instructions; and a keyboard which 'speaks' and allows the user to control the position and speed of the scanner, and the tone and volume of the speech. The words are interpreted and spoken by a voice synthesizer. At the present time, reading machines are expensive and therefore not generally available. □ see BLIND AIDS

read-only memory *n* (abbr **ROM**) a type of computer memory which may be regarded as the 'reference library' of the computer world since it holds data permanently. All microcomputers have an integrated circuit ROM which stores instructions such as the language and graphics symbols the computer uses. A typical ROM has 8 kilobytes of memory and therefore stores 65 536 bits of data on a chip contained in a 28 pin dual-in-line package. Through these pins the computer is able to select and read any one of 8192 locations in which data is stored. There are several different types of ROM. A programmable ROM (PROM) has an empty memory when first made, but it is later filled with instructions for a dedicated application (eg controlling the operations of a washing machine). An erasable programmable ROM (EPROM) can be reprogrammed with fresh instructions at any time by first erasing the old instructions with a beam of ultraviolet light. Two other types of ROM are the electrically-alterable ROM (EAROM) and the electrically-erasable programmable ROM (EEPROM) which can be reprogrammed electrically. All these types of ROM have the advantage, compared with disk and tape, that information stored in them is not lost when the power supply to them is switched off. Also, they are smaller, lighter, and more rugged than tape – compare RANDOM ACCESS MEMORY □ see EPROM, ROM CARTRIDGE, FIRMWARE, MAGNETIC BUBBLE MEMORY, PROM BURNER, FAMOS

readout *n* the display of a measurement on a meter or paper printout <*a temperature* ~>

read/write head *n* a device for transferring data to (write) and from (read) a magnetic disk in a disk drive unit □ see DISK DRIVE, MAGNETIC BUBBLE MEMORY

real time clock *n* a device for providing a computer with timing signals which are independent of the microprocessor's own timing signals. The real time clock may be used in a program to indicate the actual time of day or to control operations in real time □ see REAL TIME PROGRAMMING

real time programming *n* any computer-controlled activity where the program makes immediate decisions on events as they arise <~ *of a robot*> □ see REAL TIME CLOCK

real world *n* the environment outside a computer system <*interfacing with the* ~> □ see PERIPHERAL

receiver *n* **1** that part of a telephone containing the mouthpiece and earpiece **2** that part of a communications system which receives, decodes, and makes available to a user the information transmitted to it <*a radio* ~> – compare TRANSMITTER □ see COMMUNICATIONS SYSTEM

reception *n* the receiving of radio or television signals □ see RECEIVER

rechargeable *adj* storing electricity more than once <*a* ~ *battery*>

¹**record** *vb* to store information on magnetic tape or disk, or another storage medium, for future use <*to* ~ *a program*>

²**record** *n* **1** something (eg a program) captured on magnetic tape or disk **2** a plastic disk on which audio information is stored as minute undulations in a spiral track

recorder *n* any of a number of systems for playing and/or recording audio or video programmes □ see RECORD PLAYER, VIDEO CASSETTE RECORDER, COMPACT DISK, VIDEO DISK PLAYER

recording bias *n* the high frequency (about 60 kilohertz) alternating current fed to the recording head of a tape recorder to 'shake up' the molecular magnets on the tape which helps the tape to record sounds better □ see MAGNETIC TAPE

recording head *n* – see HEAD

record player *n* a system for recreating sounds stored on vinyl disks that consists of a motor-driven turntable, pickup, and, in some systems, a built-in amplifier and associated volume and tone controls □ see VIDEO CASSETTE RECORDER

rectifier *n* a device which changes alternating current into direct current □ see VOLTAGE REGULATOR

rectifier diode *n* a two-terminal semiconductor device or valve used in power supply circuits to change alternating current into direct current. Rectifier diodes are used in most mains-operated equipment (eg microcomputers and TVs). □ see BRIDGE RECTIFIER

redundancy *n* **1** the addition of bits to a stream of data to assist in error detection and correction □ see PARITY CHECK, HAMMING CODE **2** the building-in of duplicated hardware in a computer or other electronic system for backup purposes in the event of the failure of some function – **redundant** *adj*

reed switch *n* a fast-acting switch consisting of two metal contacts which close or open in the presence of a magnetic field. The contacts are at the ends of thin strips of easily magnetizable and demagnetizable material sealed in a glass tube containing nitrogen to reduce corrosion of the contacts. The magnetic field, from a current-carrying coil or a permanent magnet, may either close or open the contacts depending on the design of the switch. The reed switch is most useful as a remotely activated switch and is used to provide input signals to many types of electronic equipment, from telephone switching systems to robotic devices. □ see HALL-EFFECT SWITCH

reference voltage *n* a steady and unvarying voltage applied at some point in a circuit to ensure it behaves properly and consistently □ see ZENER DIODE

refraction *n* the bending of light when it passes from one transparent substance to another <~ *by a lens*>

refractive index *n* a measure of the amount of bending a beam of light undergoes in passing from one transparent substance to another. An optical fibre used for transmitting laser light in optical communications systems is designed to have a refractive index which gradually reduces from its axis to its circumference to ensure that the light does not escape from its edges. □ see OPTICAL FIBRE, REFRACTION, TOTAL INTERNAL REFLECTION

refresh *vb* **1** to send periodic bursts of signals from a microprocessor to a dynamic random-access memory to 'remind' the memory that its data is to be retained **2** to regenerate the graphics on some types of VDU – **refresh** *adj* □ see RANDOM-ACCESS MEMORY, VOLATILE

register *n* any of a number of places in a microprocessor which hold data in 8-bit, 16-bit or 32-bit sized words depending on the type of microprocessor. The data is held temporarily before being transferred to memory or being used to help carry out arithmetical and control operations. □ see ACCUMULATOR, FLAG REGISTER, INDEX REGISTER, PROGRAM COUNTER

regulate *vb* to keep the size of a current or voltage at a required value <*to ~ the voltage to a TTL logic gate*> – **regulation** *n* □ see VOLTAGE REGULATOR

regulator *n* – see VOLTAGE REGULATOR

relaxation oscillator *n* an electronic oscillator which produces a regular series of square or sawtooth waves. Its action depends on the charging of a capacitor followed by a period of 'relaxation' when the capacitor discharges through a resistor. Relaxation oscillators are used in cathode-ray tubes, in electronic musical instruments, and as electronic clocks in computing and other digital circuits. □ see ASTABLE

¹relay *vb* to receive a signal and send it on to another destination □ see TRANSPONDER

²relay *n* a magnetically operated switch which enables a small current to control a much larger current. Relays are still widely used in electronics despite the challenge from solid-state switches such as thyristors. They are especially useful for enabling computers to control motors, mains-operated equipment, and other high power devices, for they offer complete electrical isolation between the computer and these peripheral devices – compare THYRISTOR

relay station *n* a system of aerials and electronic equipment on Earth or in space for receiving radio and TV signals from one place and beaming then to another. Relay stations are used for outside broadcasts on a TV network, for transmission of telephone signals across country, and for global communications using artificial satellites. □ see TRANSPONDER, REPEATER, DIRECT BROADCAST SATELLITE, EARTH STATION

release *n* – see ADSR

reliability *n* the probability that an electronic device, circuit, or system will operate without failure <*a ~ of 99%*> – **reliable** *adj*, **reliably** *adv*

relief bush *n* a plastic device which is fitted into a cable exit hole in equipment to clamp and release the strain on a cable passing through it □ see GROMMET

relocation *n* the process of moving part of a program from one set of memory locations to another, usu to make more efficient use of memory space – **relocate** *vb*

remote access *n* any method for using a computer system located some distance away from the user <*Viewdata has ~ to a computer*> □ see TERMINAL 1

remote control *n* command over an operation (eg space exploration) by means of signals sent from a distance. The signals may be sent by wires, sound waves, light (esp infrared as in TV remote control), or by radio <*~ of model aircraft*>.

repeatability *n* the capability of a component or system to operate the same way every time it is called upon to do so <*the ~ of measurements on a meter*> – **repeatable** *adj*

repeater *n* **1** a device fitted into a long distance communications system (eg a submarine cable) to strengthen a weakening signal **2** a station in a microwave communications system that receives, strengthens, and retransmits radio signals □ see TRANSPONDER

replay *vb* to play back a recording on magnetic tape or disk for previewing and/or to edit recorded information – **replay** *n*

rerun *vb* to execute a program again – **rerun** *n* □ see EXECUTE, REPLAY

reservoir capacitor *n* also **smoothing capacitor** a large-value electrolytic capacitor used for smoothing variations in half- or full-wave rectified voltages in a DC power supply obtained from the mains AC supply □ see RECTIFIER, SMOOTHING CIRCUIT

reset *vb* to return a device or the values of a device to an original

150

state <*to ~ a binary counter*>

reset button *n* a push-button switch for setting the condition of a circuit or device to its normal or original state

resident *adj, of software* being in a computer's main memory <*a ~ control program*>

resist *n* PHOTORESIST

resistance *n* the opposition offered by a component, esp a resistor, to the flow of electricity through it □ see OHM, OHM'S LAW, REACTANCE, IMPEDANCE

resistivity *n* (a measure of) the tendency of a substance to oppose the flow of an electric current through it – compare CONDUCTIVITY

resistor *n* a component in a circuit which offers resistance to the flow of electrical current. Two or more resistors connected in series are commonly used in circuit design to divide a potential difference into a suitable ratio for controlling switching devices such as transistors. Resistors are available as discrete devices as well as being an integral part of most circuits on silicon chips. □ see VOLTAGE DIVIDER, POTENTIOMETER, RESISTOR COLOUR CODE

resistor colour code *n* a code of coloured bands printed on resistors corresponding to their values in ohms. The colours are black, brown, red, orange, yellow, green, blue, violet, grey, and white: bands of gold, silver, and red indicate the tolerance of the value. □ see PREFERRED VALUES, TOLERANCE

resistor-transistor logic *n* (abbr **RTL**) an obsolete type of digital logic circuit comprising resistors and a bipolar transistor □ see TRANSISTOR-TRANSISTOR LOGIC, COMPLEMENTARY METAL-OXIDE SEMICONDUCTOR LOGIC, EMITTER-COUPLED LOGIC

resolution *n* **1** a measure of the detail which can be obtained in images on a VDU. The maximum resolution is determined by the number of individual picture elements which make up the image. This number is about 230 thousand on a 625 line TV. A microcomputer giving high resolution images is able to control each individual picture element, while low resolution graphics occurs if the microcomputer can only control blocks of these elements. □ see PIXEL, MONITOR **2** the accuracy with which an analogue-to-digital converter (ADC) produces a binary equivalent of an analogue voltage. For an ADC which produces an 8-bit binary word, the resolution is equal to 1 part in 256 or about 0.25 per cent. □ see ANALOGUE-TO-DIGITAL CONVERTER

resonance *n* the buildup of large amplitude vibrations in an electrical or mechanical system by stimulating the system with a vibration close or nearly equal to the natural frequency of vibration of the system. Resonance is the operating principle of tuned circuits, lasers, and some types of medical diagnostic equipment. □ see TUNED CIRCUIT, LASER

resonate *vb* to make a circuit oscillate at its natural frequency of vibration □ see RESONANCE

response time *n* the time taken for a cause to have an effect (eg the time taken for a character to appear on a VDU after a key has been depressed on a computer keyboard)

retrieve *vb* to recover data from a computer memory *<to ~ a program from a floppy disk>*

return *n* an instruction in Basic at the end of a subroutine which returns control of the computer to the next instruction after the one which called the subroutine □ see RETURN ADDRESS, JUMP

return address *n* the address to which a program returns after a subroutine has been executed

reverse bias *n* a voltage applied across a pn junction to prevent the flow of electrons across the junction. The reverse bias causes a depletion region to be formed which is essential to the operation of the rectifier diode and the bipolar transistor – compare FORWARD BIAS – **reverse-biased** *adj* □ see PN JUNCTION, DEPLETION REGION

reverse video *n* also **inverse video** a facility available on some VDUs, and usu available as a command on some microcomputers, which enables the normal effect of light characters on a dark background to be reversed for emphasis or special effects

RF – see RADIO FREQUENCY

RF modulator *n* a device built into most microcomputers which imitates the signals produced by a television aerial so that an ordinary TV can be used to display program listings and graphics □ see MONITOR, RGB GUNS

RGB [*r*ed, *g*reen and *b*lue] – see RGB GUNS

RGB guns *n* the three electron guns in a colour TV or VDU tube which activate the red, green, and blue phosphor dots on the screen to enable colour images to be obtained □ see RGB MONITOR, COLOUR TELEVISION, PRIMARY COLOURS

RGB monitor *n* a VDU for displaying computer graphics in which the red, green, and blue electron guns are switched on and off directly by the computer. An RGB monitor does not require the use of modulator and demodulator circuits so it produces far better quality graphics than is possible with a normal TV receiver. □ see UHF MODULATOR, TELEVISION, COLOUR TELEVISION

ribbon cable *n* a flat and flexible lead consisting of a number of plastic-coated wires connected side by side. Peripherals (eg printers) connected to microcomputers use ribbon cable, esp where a number of signals need to be routed between devices. □ see RIBBON CONNECTOR

ribbon connector *n* a connector attached to the end of a ribbon cable □ see RIBBON CABLE

ring modulation *n* a method used by some microcomputers for making a synthesized voice by changing the amplitude of one voice pattern according to the signals of another □ see MODULATE

ring network *n* a communications system between computers in which the computers are joined together in a continuous loop. Ring networks are not very popular since data might have to pass through as many as half the computers before reaching a particular destination – compare STAR NETWORK, BUS NETWORK □ see LOCAL AREA NETWORK

ripple counter *n* a binary counter used in digital counting circuits (eg clocks) that consists of a set of flip-flops which change state one after the other – compare SYNCHRONOUS COUNTER □ see BINARY COUNTER

rise time *n* the time taken for a signal waveform to rise from 10% to 90% of its maximum voltage. A knowledge of rise time is important in digital logic circuits, which are generally designed to be activated by fast-changing signals, and in the design of filters – compare FALL TIME

RMS [*root mean square*] – see ROOT MEAN SQUARE VALUE

robot *n* **1** a computer-operated machine that can be programmed to perform repetitive, boring, and possible hazardous tasks such as paint spraying and welding on a car assembly line **2** a programmable controller (eg a numerically controlled machine tool) that accurately performs and monitors machine operations **3** an automated system for handling goods and materials in warehouses and on production lines **4** an artificial satellite or an interplanetary space probe which is designed to carry out measurements of the local environment and transmit findings back to Earth **5** a (fictional) humanoid machine that walks and talks □ see ARTIFICIAL INTELLIGENCE, AUTOMATION, ROBOTICS, ANDROID

robot arm *n* a jointed mechanical device driven by stepping motors, solenoids, or other electromechanical devices, which is able to manipulate objects (eg chess pieces) under computer control □ see AUTOMATION, ROBOTICS

robotics *n* the use of computer-controlled machines to carry out a predetermined set of operations (eg welding metal on a car production line or analysing the nature of the soil on the planet Mars) □ see ARTIFICIAL INTELLIGENCE, ROBOT, CYBERNETICS

robotize *vb* **1** to control a device or system with the aid of robots **2** to equip with robots <*the car factory was ~d*> – **robotization** *n* □ see ROBOT, ROBOTICS

roll paper *n* printer paper which is dispensed from a roll – compare FANFOLD PAPER

ROM – see READ-ONLY MEMORY

ROM cartridge *n* a read-only memory containing a program which is packaged so that it can be plugged into a microcomputer □ see FIRMWARE

root mean square value *n* also **effective value** the steady current or voltage which would give the same heating effect as an alternating current. The mains voltage in the home (240 volts AC) is the effective value of the alternating current. □ see PEAK VOLTAGE

rotary switch *n* a switch with a spindle which can be turned to select a different function or circuit □ see SWITCH

rotor *n* the movable part of a stepping motor or generator which is used to rotate cogs, wheels, etc connected to its axle – compare STATOR

route *n* any path along which information travels – **route** *vb*

routine *n* a set of computer instructions which has a specified job to do <*a ~ to generate a graphics character*> □ see LISTING, SUBROUTINE, NESTED LOOP, PROGRAM

routing *n* any procedure for allocating lines to telephone calls in a message switching system

row *n* a horizontal arrangement of data – compare COLUMN

RS flip-flop *also* **RS latch, SR flip-flop** *n* a simple flip-flop having a 'set' input and a 'reset' input. The RS latch consists of two cross-coupled NAND or NOR gates and stores a binary digit controlled by the signals applied to the two inputs. □ see FLIP-FLOP, LATCH

RS latch *n* RS FLIP-FLOP

RS-232 interface *n* a standardized method for connecting computers to peripheral equipment such as printers. Bits of digital data are sent one after the other (serially) down the connecting line – compare IEEE-488 INTERFACE □ see SERIAL INTERFACE

RTL – see RESISTOR-TRANSISTOR LOGIC

RTS [*ready to send*] a signal from a computer peripheral to indicate that it has data to send

rubber banding *n* a method used in computer graphics of stretching and pulling out lines attached to a common vertex without moving their other ends

rumble *n* a low frequency mechanical noise produced by the bearings of poor quality hi-fi turntables. Rumble is a mechanical fault which is not present in modern compact disk record players. □ see COMPACT DISK

run *vb* to execute a program – **run** *n* □ see EXECUTE

runtime *n* the time needed to run a program □ see RUN

S

s the symbol for a second of time

S the symbol for the unit of electrical conductivity, the siemen □ see SIEMEN

S100 *n* a connecting system for computers comprising 100 pins and designed for Intel 8080 and similar microprocessors □ see RS-232 INTERFACE, IEEE-488 INTERFACE

sample *vb* to measure the value of an electrical quantity *<to ~ the amplitude of a signal>* – **sample** *n* □ see DIGITAL RECORDING

sample and hold *n* (abbr **S/H**) a circuit based on a capacitor that charges up and holds the maximum value of a signal. A sample and hold circuit is the basic element of analogue-to-digital converters used in digital recording and speech recognition systems. □ see DIGITAL RECORDING, SPEECH RECOGNITION

sampling rate *n* the number of times per second that the value of a quantity is measured □ see PULSE CODE MODULATION, COMPACT DISK, DIGITAL RECORDING

satellite *n* □ see COMMUNICATIONS SATELLITE, VOYAGER, VIKING, INTELSAT

saturation *n* the state of an active device (eg an operational amplifier) when its output signal reaches a maximum value despite further increases in the strength of its input signal. Saturation is an undesirable state in an amplifier but is necessary in Schmitt triggers and other solid state switching devices – **saturate** *vb*

save *vb* to store programs on magnetic tape, disk, or EPROM by transferring them from a random-access memory inside a microcomputer – compare LOAD

scalar *adj* represented by a size only *<temperature is a ~ quantity>* – compare VECTOR – **scalar** *n* □ see ANALOGUE

scale *n* **1** something having markings indicating degrees or quantity that allow a measurement to be taken *<a tuning ~ on a radio>* **2** an instrument or machine for weighing items *<a digital ~>*

scale-of-ten counter *n* DECADE COUNTER

scaling *n* the process of producing a set of values fitting into a specified range – **scale** *vb*

scan *vb* to move a beam of laser light, electrons, etc repeatedly over an object, screen, etc for the purpose of examination or to produce an image □ see SCANNING OPTICAL MICROSCOPE, SCANNING ELECTRON MICROSCOPE, RASTER SCANNING, ULTRASONIC SCANNER, BAR CODING

scan frequency *n* SCAN RATE

scanner *n* any device for directing a beam of sound, light, radio waves, or electrons at a target for examination purposes or to produce an image

scanning beam *n* a beam of electrons in a TV or video camera which scans the image back and forth and creates an electrical current representing the light variations in the TV scene □ see TELEVISION, VIDEO CAMERA

scanning electron microscope *n* a high-power microscope which uses a fine beam of electrons to repeatedly scan an object in a vacuum to produce a detailed image on the screen of a cathode-ray tube. The scanning electron microscope is often used to study the fine detail of an integrated circuit, esp in the development of new microelectronics devices on silicon chips – compare SCANNING OPTICAL MICROSCOPE

scanning optical microscope *n* (abbr **SOM**) an instrument for the detailed examination of a silicon chip using a finely focussed beam of laser light which is moved back and forth across the surface of the chip in a series of fine lines. The light reflected from the surface of the chip is detected by a photodiode and the electrical signal is processed to form an image on a VDU. Although the scanning optical microscope does not produce such a highly magnified image as a scanning electron microscope, it is simpler to operate since it does not require a vacuum and the laser beam is less damaging than electrons to the surface of the chip – compare SCANNING ELECTRON MICROSCOPE

scan rate *also* **scan frequency** *n* the number of scans carried out per second □ see RASTER SCANNING

schematic *n* a circuit diagram which shows the important, but not all, connections required for a particular application <*an interfacing* ~> – **schematic** *adj* □ see BLOCK DIAGRAM

Schmitt trigger *n* a snap-action electronic switch which goes off and on at two specific input voltages, the upper and lower threshold voltages. The Schmitt trigger is widely used for 'sharpening up' slowly changing waveforms, and for eliminating noise from circuits. Schmitt triggers are available in integrated circuit packages, and they are so useful in circuit design that a single IC contains up to six Schmitt triggers. □ see THRESHOLD, HYSTERESIS, POSITIVE FEEDBACK

Schottky diode *n* a special diode used in some types of transistor-transistor logic (TTL) digital integrated circuits to speed up the operation of the bipolar transistors on the chip □ see TRANSISTOR-TRANSISTOR LOGIC

scintillation *n* the flash or flashes of light produced by certain phospors or crystals which have absorbed a single electron or proton, or which have been struck by a quantum of gamma or X-ray radiation. Scintillation is responsible for the light emitted from the phosphor on the screen of a cathode-ray tube in a TV or VDU. □ see PHOSPHOR, CATHODE-RAY TUBE, SCINTILLATION COUNTER

scintillation counter *n* an electronic instrument for measuring the amount of radiation given off from a radioactive source. A scintillation counter counts the flashes of light produced by the absorption of the radiation by a phosphor or a crystal. □ see GAMMA COUNTER

scissoring *n* a way of removing parts of a display which lie outside of a window □ see WINDOW

scope *n* [oscillo*scope*] – see CATHODE-RAY OSCILLOSCOPE

SCR – see SILICON-CONTROLLED RECTIFIER

scramble *vb* to transmit a message in a form which makes it difficult to understand unless it is received by a device specially equipped to decode it

scratch *vb* to erase some data from a memory device

scratchpad *n* a small booklet of blank copies of a breadboard layout for use in sketching the design of circuits □ see BREADBOARD

scratchpad memory *n* an area in computer memory where data is stored temporarily and where it can be accessed quickly when required

screen *n* **1** the front part of a VDU, TV, or other display device, on which visual information is displayed □ see CATHODE-RAY TUBE, VDU, PHOSPHOR, LIQUID CRYSTAL DISPLAY **2** any device (eg a metal case), usu connected to earth, which shields equipment or a device from electrical noise □ see COAXIAL CABLE, SLEEVING

screened cable *n* a wire or group of wires wrapped in an earthed metal skin. The screen reduces interference to messages carried by the inner wire T □ see COAXIAL CABLE

screen glare *n* unwanted reflections from the screen of a VDU. Screen glare causes eyestrain, stress, and tiredness and is usu reduced by a special filter (eg a polarizing filter) placed over the screen. □ see POLARIZING FILTER

screen memory *n* also **display file** a part of the memory of a microcomputer consisting of a set of consecutive memory locations which are used to store information to be displayed on the screen. Each memory location contains a unit (usu a byte) of data representing a single position on the screen. By changing the contents of each location, the programmer can change the shape of the characters displayed.

scrolling *n* the process of moving information, esp text, upwards or sideways on a VDU to make room for new information – **scroll** *vb*

search *vb* to look through a set of items (eg files on a disk) to find an item required – **search** *n*

secondary cell *n* an electric cell which can be recharged after use – compare PRIMARY CELL

secondary coil *also* **secondary winding** *n* the winding on a transformer which supplies the output voltage – compare PRIMARY COIL □ see TRANSFORMER

secondary colour *n* a colour produced by mixing together two of the primary colours, red, green, and blue, which is the technique used for displaying a wide range of colours on a VDU □ see PRIMARY COLOURS, COLOUR TELEVISION

secondary storage *n* BACKING STORE
second generation *adj* – see COMPUTER GENERATIONS
second sourcing *n* the practice whereby one semiconductor
manufacturer arranges for another manufacturer to make one of its
devices under licence to ensure continued supply in the event that
the first manufacturer does not have enough capacity to produce the
device <~ *the Z80 microprocessor chip*>
sector *n* a subdivision of a track on the surface of a magnetic disk
□ see TRACK, SOFT-SECTORED DISK, HARD-SECTORED DISK,
FORMATTING
security code *n* a code enabling someone to have private access to
information stored in a computer system □ see PIN, SCRAMBLE
security electronics *n* the use of electronic devices (eg ultrasonic
transducers) and systems (eg radio) for protecting people and places
from interference and damage □ see INTRUDER ALARM, PIN, SMART
CARD
seed crystal *n* a small pure crystal of silicon on which a much larger
crystal grows when it is immersed in a vessel containing molten
silicon. The growing crystal is rotated and pulled out of the vessel
slowly to produce a boule from which silicon chips are made. Large
pure crystals of germanium and gallium arsenide are grown in the
same way. □ see BOULE, WAFER, SILICON CHIP
segment *n* a short barlike part of an alphanumeric charcter on a
seven-segment display
seismic survey *n* the use of mechanical energy waves to find out
about the geological structure extending hundreds of feet below the
ground. The usual purpose of a seismic survey is to look for possible
petroleum or gas deposits. The energy waves are often produced by
specially built vehicles equipped with vibrator pads which are
lowered to the surface of a road. The vibrations generated by these
pads are reflected by underground rock layers and picked up by a
string of geophones. The sound pattern picked up by these small and
sensitive microphones is recorded in digital form on tape in another
special vehicle and processed later by a computer.
seismograph *n* an electronic instrument for measuring the strength
of earth tremors □ see SEISMIC SURVEY
selective *adj* responding to a narrow band of audio or radio
frequencies <*a ~ tuned circuit in a radio receiver*> □ see Q-FACTOR,
NOTCH FILTER
selectivity *n* the ability of a tuned circuit in a radio receiver to
separate closely-spaced stations on a broadcast waveband – compare
SENSITIVITY
semiconductor *n* **1** a solid material which conducts electricity better
than an insulator (eg glass) but not as well as a metal (eg copper).
The majority of microelectronic devices are based on the
semiconductor silicon which has largely replaced germanium.
Diodes, transistors and integrated circuits are the main devices made

from semiconductors, and these are used as switches, amplifiers, counters, and timers in all sorts of consumer and industrial equipment ranging from robots to wristwatches, and from wordprocessors to weather satellites. All these applications are possible because silicon can have its electrical properties modified dramatically and usefully by adding small amounts of impurities. □ see N-TYPE, P-TYPE, PN JUNCTION **2** *also* **semiconductor device** an electronic component containing one or more solid-state elements (eg transistors) and terminals for connecting it into a circuit □ see BIPOLAR TRANSISTOR, FIELD-EFFECTIVE TRANSISTOR, TRIAC, INTEGRATED CIRCUIT

semiconductor physics *n* a branch of physics concerned with explaining and predicting the behaviour of semiconductors (eg silicon and gallium arsenide), esp with regard to their usefulness for electronic applications □ see ELECTRONICS, QUANTUM THEORY

sensitivity *n* **1** the ability of a radio receiver to detect weak signals at its aerial input – compare SELECTIVITY **2** the ability of an analogue meter, esp a multimeter, to measure voltage while drawing the least possible current from a circuit. Sensitivity is expressed in units of ohms per volt, and the greater its value the greater the sensitivity of the meter (eg a 20 000 ohms per volt meter draws a current of 50 microamperes). – **sensitive** *adj*

sensor *n* a device which detects changes (eg of pressure) in its surroundings and outputs an electrical or other signal – **sense** *vb* □ see TRANSDUCER, STRAIN GAUGE, THERMOCOUPLE

sequence *n* a series of items in (numerical) order <*a ~ of addresses*> – **sequence** *vb* □ see SEQUENTIAL PROCESSING

sequential-access *adj* also **serial access** *of a file, memory device, etc* containing data that can only be accessed in the sequence in which it was stored – compare RANDOM-ACCESS

sequential logic *n* a digital logic circuit (eg a binary counter) which can store information received from previous combinations of inputs, ie it has a memory. Sequential logic produces an output combination of logic 1s and 0s which is determined not just on the present set of 1s and 0s, but also on previous combinations of binary signals at its inputs – compare COMBINATIONAL LOGIC □ see FLIP-FLOP, BINARY COUNTER, SHIFT REGISTER

sequential processing *n* the processing by a computer of a number of items sequentially (ie one after the other) – compare PARALLEL PROCESSING

serial *adj, of events or processes* occurring one after the other <*~ processing in a microprocessor*> – **serially** *adv* □ see SEQUENTIAL LOGIC

serial access *n* SEQUENTIAL ACCESS

serial input-output *n* a method of transferring data one bit at a time between a computer and a peripheral device (eg a printer) – compare PARALLEL INPUT-OUTPUT

serial interface *n* a circuit connecting a computer with a peripheral (eg a printer) through which data is transmitted one bit at a time – compare PARALLEL INTERFACE □ see UART, USART, SERIAL SIGNALS, RS-232 INTERFACE

serial signals *n* electrical signals which travel through a circuit one after the other. A program on tape enters a microcomputer as serial signals – compare PARALLEL SIGNALS

series circuit *n* a circuit in which components (eg resistors) are connected end-to-end so that the same current flows through each component – compare PARALLEL CIRCUIT

servomechanism *n* also **actuator** a device (eg a power steering mechanism) which controls large amounts of mechanical power by means of small amounts of mechanical power. Some industrial robots use servomechanisms. □ see SERVOSYSTEM

servosystem *n* an electromechanical system (eg the device controlling the movement of the read/write head on a disk drive) which is controlled precisely through the use of sensors which monitor its movement and provide control signals. □ see COMPACT DISK, VIDEO DISK

seven-segment display *n* a device for displaying numbers and some letters by making visible selected combinations of seven segments arranged in the form of a figure 8. The seven-segment display is widely used in digital instruments (eg watches). There are two common types of seven-segment display, the liquid crystal display and the light-emitting diode display. □ see LIGHT-EMITTING DIODE, LIQUID CRYSTAL DISPLAY

S/H – see SAMPLE AND HOLD

shadow mask tube *n* a type of cathode-ray tube used in colour television receivers in which a metal mask with holes in it ensures that the three electron beams reach the corresponding red, green, and blue phosphor dots on the screen □ see COLOUR TELEVISION, RGB GUNS

shared logic facility *n* a computer system having a dispersed collection of terminals (eg wordprocessors) all of which have access to one microprocessor and its storage devices □ see TERMINAL, TIMESHARING

shelf life *n* the length of time that a product, esp a cell or battery, may be kept in store without deteriorating or weakening □ see LITHIUM CELL

shift register *n* a string of transistor flip-flops contained in an integrated circuit package that hold binary data which can be moved, bit by bit, right or left through the flip-flops. The shifting of the bits is done with clock pulses, and the bits can be output from the shift register using a 'read' signal. Shift registers are used in arithmetic units in computers for multiplication and division of binary words, and in I/O circuits to convert serial data to parallel data. □ see REGISTER, COUNTER

Shockley, William – see BARDEEN

short *n* SHORT CIRCUIT

short-circuit *also* **short** *n* an accidental connection between two parts of a circuit, usu the result of a path of low resistance to current flow. Circuits are generally protected against a short-circuit by a fuse which 'blows' when excess current flows through it. – **short-circuit** *vb* □ see FUSE

short waves *n* radio waves which have wavelengths between about 120 metres and 20 metres (ie with frequencies between 2.5 megahertz and 15 megahertz respectively), and which are used mainly for amateur and long-range communications □ see MICROWAVES, MEDIUM WAVES, LONG WAVES

shunt *vb* to connect one component in parallel with another so as to reduce the current flow through the first component – **shunt** *n*

SID [*s*ound *i*nterface *d*evice] a sophisticated integrated circuit incorporated in some microcomputers for generating sound effects. The chip can produce three separate frequencies, each with an eight-octave range and volume control, four separate waveforms, filters, and ADSR envelope control. □ see ADSR, SPEECH SYNTHESIZER

sidebands *n* the two radio frequencies, one slightly above and one slightly below the carrier frequency, which are produced when a radio frequency (carrier) wave is amplitude-modulated by an audio frequency wave. Each audio frequency wave has its own two sidebands so that a band of frequencies is produced called the bandwidth of the radio signal. Sidebands are also produced when a radio frequency carrier wave is frequency modulated. □ see AMPLITUDE MODULATION, FREQUENCY MODULATION, BANDWIDTH

siemen *n* the unit of electrical conductance, equal to the reciprocal of the ohm, ie 1/ohm. Hence the greater the resistance of a material, the lower the conductance. The siemen is used to compare the electrical properties of materials (eg the conductance of copper is greater than that of silicon). □ see CONDUCTIVITY

sign *n* a symbol (+ or −) which indicates whether a number is positive or negative. A number without a sign is assumed to be positive. □ see POSITIVE, NEGATIVE

signal *n* any message transmitted from one place to another <*a Morse code* ~> – **signal** *vb* □ see COMMUNICATIONS SYSTEM

signal generator *n* WAVEFORM GENERATOR

signal shaping *n* the use of components and circuits to change the shape of a signal in a useful way (eg 'squaring' a signal using a Schmitt trigger) □ see SCHMITT TRIGGER, WAVEFORM

signal-to-noise ratio *also* **S/N ratio** (abbr **SNR**) *n* the ratio of the strength of a signal to the strength of any background noise which might also be present

sign bit *n* a 0 or 1 in front of a binary word to indicate the sign of the number. A 1 bit means the number is negative, a 0 bit that it is positive

SIL – see SINGLE-IN-LINE
silica *n* SILICON DIOXIDE
silicon *n* an abundant nonmetallic element of atomic number 14
which has largely replaced germanium as the basis of diodes,
transistors, and integrated circuits. Like germanium, silicon forms
crystals in which neighbouring atoms link together by sharing
electrons, a process called 'covalent bonding'. A pure crystal of
silicon can have its electrical properties modified usefully to produce
p-type and n-type semiconductors. Silicon is more abundant than
germanium since it is combined with oxygen in sand, and it has the
advantage, compared with germanium-based semiconductors, that
semiconductor devices produced from silicon are less likely to suffer
damage from self-heating (or thermal runaway) – compare
GERMANIUM ☐ see P-TYPE, N-TYPE, PN JUNCTION, BIPOLAR
TRANSISTOR, VALENCY
silicon chip *n* a small piece of silicon on which complex miniaturized
circuits are made by photographic and chemical processes. The
silicon chip is cheap to make in large quantities, it can withstand
rough treatment, it is reliable in operation, and it needs very little
electrical power; therefore it has become the heart of computer,
control, and communications systems. ☐ see WAFER,
MICROELECTRONICS, PHOTOLITHOGRAPHY, VERY LARGE SCALE
INTEGRATION
silicon-controlled rectifier *n* (abbr **SCR**) also **thyristor** a three-
terminal semiconductor rectifier designed to control the power
supplied to a load. A small positive potential difference applied
between the gate and cathode terminals allows a large current to
flow between the anode and cathode terminals. This current
continues to flow even when the gate voltage is removed. In the
control of AC power (eg in a wall-mounted light dimmer) the silicon-
controlled rectifier conducts for only a part of each cycle, the time
during which conduction occurs being controlled by a variable
resistor and capacitor. ☐ see TRIAC, DIMMER
silicon dioxide *n* also **silica** an insulating material laid down as a
thin skin on the surface of a pure silicon wafer by heating the wafer
in pure oxygen. Impurities of different types are allowed to diffuse
through holes etched in the silicon dioxide skin into the silicon to
produce transistors and other semiconductor devices making up a
circuit on a silicon chip. ☐ see PHOTOLITHOGRAPHY, GATE OXIDE
silicone *n* a synthetic substance based on silicon but not used as the
basis for semiconductor devices. Silicone is usu applied as a
waterproof grease to protect components and hardware from the
effects of moisture. ☐ see SILICON
Silicon Glen an area in the West Lothian of Scotland where a
number of manufacturers and users of high technology devices, esp
semiconductors, are encouraged to set up business ☐ see SILICON
VALLEY

silicon nitride *n* an electrically insulating substance used (1) for storing electrical charge in some types of integrated circuits (eg MNOS devices) (2) for resisting the attack of etchants in the manufacture of silicon chips, so enabling fine detail to be defined through gaps in a layer of silicon dioxide □ see MNOS, PHOTOLITHOGRAPHY

silicon-on-sapphire *n* (abbr **SOS**) a technology for making metal-oxide semiconductor integrated circuits in which transistors are built up in thin layers on an inert substrate of synthetic sapphire. The technique produces integrated circuits which are faster in operation than conventional monolithic integrated circuits. □ see SUBSTRATE

Silicon Valley an area near Sunnyvale to the south of San Francisco, California, where many computer and semiconductor manufacturers are situated □ see SILICON GLEN

silicon wafer *n* WAFER

simulation *n* (the use of) a model or 'make believe' of a real system using a computer and one or more VDUs for displaying computer-generated or prerecorded pictures. Simulation is widely used for training airline pilots and astronauts. – **simulate** *vb,* **simulator** *n*

Sinclair, Clive. Founder of Sinclair Radionics (1962) which produced the first pocket calculator in 1972 and confirmed his flair for miniaturizing products and for selling them in large numbers. Sinclair founded Sinclair Research in 1979 and produced the ZX80 microcomputer followed by the ZX81, the Sinclair Spectrum, and the Quantum Leap in 1984. His company, backed by his own research centre in Cambridge, has produced a novel flat-screen TV and is developing a battery car, both products aimed at a mass market. □ see FLAT-SCREEN TV

sine wave *n* a smoothly changing waveform which is represented by the mains alternating current. Sine waves are used for testing audio and other circuits. □ see WAVEFORM GENERATOR

single-board computer *n* a microcomputer which has its microprocessor, memory, and input/output interfaces all on a single printed circuit board

single-density disk *n* – see DOUBLE-DENSITY DISK, FLOPPY DISK

single-in-line *adj* (abbr **SIL**) *of a component package* having a number of pins arranged in a single line and separated by 0.1 inch (2.54 millimetres) <*a ~ resistor package*> □ see DUAL-IN-LINE, QUAD-IN-LINE

single-sided disk *n* a floppy disk only one side of which can be recorded on – compare DOUBLE-SIDED DISK

single-step *vb* to execute a program one instruction at a time, usu by operating a switch on the computer's console – **single-step** *adj*

sink *vb* to accept a current <*the output terminal of the logic gate ~s 1.6mA*> – compare SOURCE – **sink** *n*

sinusoidal *adj* like a sine wave <*a ~ waveform*> □ see SINE WAVE, WAVEFORM GENERATOR

skin-effect *n* the tendency for high frequency signals to move along the surface of a wire rather than along its middle. Since the skin-effect increases the resistance of ordinary copper wire, radio tuning coils, for example, consist of many fine strands of copper wire. This increases the effective surface area of the wire, lowering its overall resistance to radio signals, and thereby improving the selectivity of the tuned circuit. □ see TUNED CIRCUIT, SELECTIVITY

sky wave *n* a radio wave from a transmitter which reaches a radio receiver by reflection from the ionosphere, provided the frequency of the radio wave is below about 30 megahertz. The repeated reflection of a sky wave between the ionosphere and the ground is responsible for round-the-world radio communication. □ see IONOSPHERE, GROUND WAVE

slave *adj* capable of being controlled on command <*a ~ flash unit in photography*>

sleeving *n* a covering of plastic, cotton, porcelain, or other electrically insulating material, which is slid over a bare wire to protect it against possible electrical shorts □ see HEAT-SHRINKABLE SLEEVING

slice *n* WAFER

slot *n* **1** a narrow space containing connectors which holds a removable printed circuit board **2** a space in a data storage system which is of the right size to hold a specified amount of data <*a page ~*> **3** an interval of time <*a broadcast programme ~*>

SLSI – see SUPER LARGE SCALE INTEGRATION

small scale integration *n* (abbr **SSI**) the process or technology of making integrated circuits with less than 20 logic gates on the chip □ see MOORE'S LAW

smart *adj, of an electronic device* able to make decisions on information it receives <*a ~ data logger*> □ see SMART CARD, ARTIFICIAL INTELLIGENCE

smart card *n* a credit card-sized piece of plastic containing integrated circuits which can compute and hold data. Smart cards are being used in France as 'electronic money' at point-of-sale terminals in stores and in pay telephones. The bank which issues a smart card to an account holder writes into the card's built-in memory chip the holder's account number, an agreed monthly spending limit, a personal identity number (PIN), and a secret code word associated with the bank. In a store, the card is plugged into the point-of-sale terminal which supplies power to the chips in the card. The terminal checks the amount spent against the card holder's monthly spending limit which is automatically updated by the bank. If the customer is not overspent, his or her PIN is then entered and this is checked against the PIN stored in the card's read-only memory. If the numbers agree, the terminal writes the date and amount into the customer's card, and the transaction is complete. Smart cards are being considered as a means of carrying a person's medical history and as a security pass. □ see POINT-OF-SALE TERMINAL, ELECTRONIC FUNDS TRANSFER

smoke test *n* the act of switching on a newly repaired or brand new electronic device to see if it works

smoothing capacitor *n* RESERVOIR CAPACITOR

smoothing circuit *n* a circuit consisting of one or two large-value electrolytic capacitors for smoothing half- or full-wave rectified DC voltages in mains-operated power supplies □ see RESERVOIR CAPACITOR

SNR – see SIGNAL-TO-NOISE RATIO

S/N ratio *n* SIGNAL-TO-NOISE RATIO

socket *n* a part of an electrical system into which a plug is fitted to make an electrical connection <*a mains power* ~> □ see DIN PLUG

soft copy *n* information which is not in physical form (eg flight arrivals on a VDU in an airport lounge) – compare HARD COPY

soft-sectored disk *n* a magnetic disk which is divided into tracks and sectors by the microcomputer when the disk is formatted – compare HARD-SECTORED DISK □ see FORMATTING

software *n* instructions or programs stored in and manipulated by a computer system, esp as contrasted with the electronic equipment of the system. Some software is stored in a computer's read-only memory (ROM) and performs housekeeping functions. Other software is entered into the computer's random-access memory (RAM) from an external store for particular applications (eg educational games) – compare HARDWARE □ see FIRMWARE, UTILITY, APPLICATIONS PROGRAM, DISK OPERATING SYSTEM

software engineering *n* the design and development of computer programs for particular applications

software house *n* a company specializing in writing and selling computer programs. A software house provides manuals and other documentation, and sometimes offers advice on appropriate computer systems.

solar cell *n* a device for converting the energy of sunlight or of artificial light into electricity <*a solar panel is a bank of* ~s> □ see TRANSDUCER

solder *n* an alloy of tin and lead (typically in the ratio 60:40) used for joining metals together (eg component leads to tracks on circuit boards). Solder is usu used in the form of a wire which has an inner core of flux to help clean and prepare the materials to be joined together. – **solder** *vb* □ see SOLDERING IRON

soldering iron *n* a tool consisting of an electrically heated copper end ('bit') attached to a handle and used for melting solder □ see SOLDER, PRINTED CIRCUIT BOARD

solenoid *n* a coil of copper wire wound on a tube in which an iron rod is moved by a magnetic field produced by passing current through the coil. Solenoids are used where a single or repeated push or pull action is required, as in the design of robotic systems.

solid-state *adj* relating to the crystal structure, the physical and chemical properties, and the uses of solid materials, esp silicon and other semiconductors <*a ~ amplifier*> □ see SEMICONDUCTOR PHYSICS

solid-state relay *n* a device based on a silicon-controlled rectifier for switching a large current on and off by means of a small current. Unlike an electromechanical relay, the solid-state relay has no moving parts and it is smaller, simpler, and faster in operation. However, it does not offer the complete electrical isolation between the two currents that the electromechanical relay does. □ see RELAY, SILICON-CONTROLLED RECTIFIER, TRIAC

SOM – see SCANNING OPTICAL MICROSCOPE

sort *n* a program or method used for rearranging data into a particular order – **sort** *vb*

SOS – see SILICON-ON-SAPPHIRE

sound carrier *n* a radio wave which carries audio information (eg radio broadcasts on the medium waveband) – compare VISION CARRIER □ see CARRIER WAVE

sound effects *n* sounds, not all musical and including synthesized speech, which some computers can generate to liven up a program and make it more realistic □ see SPEECH SYNTHESIS, ELECTRONIC MUSIC

sound signal *n* an electrical signal produced by a microphone or similar device

¹source *vb* to deliver a current <*the op amp ~*s *10mA*> – compare SINK – **source** *n*

²source *n* **1** the one of the three terminals of a field-effect transistor which allows current to leave the channel of the device after having moved from the drain terminal □ see DRAIN, FIELD-EFFECT TRANSISTOR **2** the place from which information is sent <*a radio ~*>

source code *n* the language in which a computer program is written <*Basic is a ~*> □ see OBJECT CODE, COMPILER

source follower *n* a field-effect transistor amplifier in which the output signal is taken from across a resistor connected in its source lead. The output signal 'follows', ie has the same sign, as the input signal applied to the gate terminal. The source follower is useful esp for matching a high impedance input transducer, such as a crystal microphone, to a conventional lower impedance amplifier stage. □ see EMITTER FOLLOWER

space charge *n* DEPLETION REGION

spacelab *n* a self-contained set of instruments designed to fit into the payload bay of the space shuttle. These instruments, which depend largely on electronic devices for sensing, amplifying, counting, etc, cover a wide range of investigations. For example, the Sun is being studied to gain a better understanding of how it produces its energy; measurements of x-ray emissions from clusters of far-off galaxies and the energies of cosmic rays are helping us understand the origin of the universe; the birth of new stars is being investigated with the help of an infrared telescope; processes in the Earth's upper atmosphere are being looked at to help in assessing the effects of atmospheric pollution; the space shuttle's crew provide blood samples to determine what makes bone weaker in weightless conditions; extremely pure materials are being produced in the weightless conditions of the spacelab which should lead to improved semiconductors; and plants and animals are being studied to help understand the effects of weightlessness on growth. □ see SPACE SHUTTLE, SPACE TELESCOPE

space shuttle n a reusable Earth-orbiting spacecraft which is launched vertically with rocket assistance and which glides back to Earth to land like a normal aircraft. The assisting rockets are jettisoned after use and fall back into the sea for reuse. The underside of the space shuttle is covered with heat-resisting tiles to withstand the heat generated by friction during re-entry into the Earth's atmosphere. The shuttle was developed to reduce the cost of manned flights into space, and to explore the possibilities of servicing and launching artificial satellites, of making extended observations of the Earth and space using on-board instruments, and of building space stations which could be assembled and manned using the space shuttle as a 'shuttle service'. The successful activities of the space shuttle are dependent on the skills and knowledge accumulated over many years of space exploration, and esp on the use of microelectronics in complex computer, communications, and instrumentation systems. □ see SPACELAB, SPACE TELESCOPE

space telescope n a multi-purpose optical telescope due for launching and servicing by the space shuttle in the late 1980s. Although the space telescope will have a smaller mirror than many Earth-based telescopes, it will be able to see seven times deeper into space, and even to the edges of the visible universe. Because the space telescope will not be affected by the poor 'seeing' caused by the Earth's atmosphere, and by the bending of the structure caused by gravity, astronomers will be able to look at quasars, galaxies, and supernovae, and search for planets round distant star systems which are 50 times fainter than those visible from Earth. The telescope will be pointed extremely accurately at these objects using on-board electronic control systems. Electronic instruments and their sensors in the focal plane of its mirror will record and store data for retrieval by subsequent space shuttle missions. □ see SPACE SHUTTLE, SPACELAB

spark printer n a printer which uses sparks to burn off the surface of aluminium-coated paper to leave the shapes of characters

speaker n LOUDSPEAKER

specification n a detailed list of the operating characteristics and facilities offered by something <the ~ of a supercomputer> – **specify** vb

spectral response *n* the variation with light strength of the electrical output of a light-sensitive electronic device (eg a solar cell)

spectrum *n* the complete range of frequencies of sound, light, or other electromagnetic radiation *<the visible ~>* □ see ELECTROMAGNETIC WAVES

spectrum analyser *n* an instrument for measuring how sounds are made up from notes of different frequencies. A spectrum analyser can be used to find out how audio amplifiers respond to different frequencies and for studying the make-up of artificial speech. □ see SPEECH SYNTHESIS

speech chip *n* an integrated circuit which provides electronically synthesized words (eg for use in computer games) □ see SPEECH SYNTHESIS

speech recognition *n* the process of getting a computer to recognize spoken words by comparing them with words stored in its memory as patterns of electrical signals. Speech recognition is a more difficult process than speech synthesis – compare SPEECH SYNTHESIS

speech synthesis *n* the generation of artificial speech using computers or purpose-designed integrated circuits. Speech synthesis makes computers more user-friendly, and the readout from many types of instrument (eg those in cars) is often easier (and safer) to hear than to see. Users of microcomputers can buy hardware and software to elaborate games programs with verbal warnings, instructions, and encouragement to the player. Speech synthesis is used in many games (eg chess), and household equipment (eg microwave cookers) is available which will provide spoken information about cooking instructions for a specific food. □ see PHONEME, ALLOHONE, SPEECH RECOGNITION, ARTIFICIAL INTELLIGENCE, MAN-MACHINE INTERFACE, TRIP COMPUTER

speed *n* the rate at which a device operates □ see PROPAGATION SPEED

spider *n* a network of metal connections inside an integrated circuit package which connects the pins to the silicon chip □ see PIN

spike *n* a sudden and short-lived burst of electrical interference, usu generated in a power supply (eg the mains), which can make a program or a computer crash □ see BUG, CRASH, GLITCH, TRANSIENT

split-screen *adj, of a VDU or monitor* able to display graphics and/ or text in more than one box (or 'window') on the screen at the same time □ see WINDOW, TOUCH-SCREEN COMPUTER

spreadsheet *n* a tabular arrangement of numerical data displayed on a VDU that allows rapid automatic calculations to be made; *also* software that provides this facility. A spreadsheet acts as the computer equivalent of a sheet of paper, a pencil, a rubber, and a calculator, and is used to help scientists, engineers, and businessmen make cost-effective use of their resources. The spreadsheet has been responsible for selling more computers to business than any number

of technical features and it is designed to be as easy to use as a
games program. It enables a user to build up an actual picture on the
VDU of his/her activities, and then see the effects of any prospective
change in policy, pricing, margins, tax, or anything else which is
relevant. □ see BUSINESS COMPUTER

sprite graphics *n* a method used by some microcomputers to
produce images on the screen of a VDU which are more realistic
than is possible with conventional graphics. Sprite graphics allows a
character to be 'drawn' on any one of a number of 'layers' on the
screen, thus enabling the computer artist to make one sprite (eg a
cloud) appear to move behind another sprite (eg a tree) giving a
three-dimensional effect. Once a character, called an 'object' has
been drawn it can be made to move across the screen in any
direction at a specified speed. □ see GRAPHICS, GRAPHIC
CHARACTERS

sprocket feed *n* TRACTOR FEED

sprockets *n* pins which fit into holes along the edge of fanfold paper
□ see FANFOLD PAPER

Sputnik the first of the Earth's artificial satellites launched by the
Soviet Union on October 4 1957. Sputnik carried the first electronic
instruments into space and measured temperature and electron
densities in the upper atmosphere. □ see EXPLORER, ARTIFICIAL
SATELLITE, TELEMETRY

square wave *n* a signal waveform having an amplitude which
changes sharply from one value (high) to another (low) giving it a
square shape when seen on an oscilloscope. The square wave is
produced by a quartz crystal clock in a computer where it is used for
timing and controlling the various operations performed by the
microprocessor. □ see CRYSTAL CLOCK, ASTABLE

SR flip-flop *n* RS FLIP-FLOP

SSI – see SMALL SCALE INTEGRATION

stabilize *vb* to make a circuit have a constant performance despite
temperature changes and other effects on components *<to ~ the
output of a power supply>* – **stabilization** *n* **stability** *n* □ see
VOLTAGE REGULATOR, ZENER DIODE

stabilizer *n* a circuit which keeps an electrical quantity, usu voltage,
at a fixed value □ see VOLTAGE REGULATOR

stable *adj* unvarying in operation *<a ~ oscillator>* □ see
STABILIZER

stable state *n* an unchanging condition of a circuit *<a flip-flop has
two ~s>* □ see ASTABLE, BISTABLE

stack *n* a section of computer memory used for storing a list of data
which is available to a user on the 'lifo' principle, ie last in, first out.
There are several types of stack used for different purposes (eg
Gosub stack, calculator stack, and machine stack).

stage *n* any part of a circuit or system which by itself, or together
with similar parts, forms part of a larger system *<a two*-stage

amplifier>

stand-alone *adj* FREESTANDING

standard interface *n* any hardware connection which conforms to an agreed standard *<RS-232 is a ~>*

standardization *n* the process of incorporating certain designs and conventions in equipment to ensure that products from different manufacturers are interchangeable

star network *n* a system which uses a central controlling computer to link several computers together and to enable them to share certain peripherals □ see RING NETWORK, BUS NETWORK

start bit *n* a bit (binary 1 or 0) which is transmitted in an asynchronous communications system to tell a receiver device that information is about to follow – compare STOP BIT □ see UART

statement *n* a self-contained instruction or set of instructions in a computer language – see INSTRUCTION, MULTISTATEMENT LINE

¹static *adj* **1** *of a computer memory* only losing stored information when the power supply to it is switched off *<~ random-access memory>* – compare DYNAMIC □ see RANDOM-ACCESS MEMORY **2** of or being electrical effects caused by charges at rest *<~ electricity>* □ see ELECTROSTATICS

²static *n* electrical interference on a communications channel caused by unusual atmospheric conditions, lightning, electrical machinery, etc *<~ from car ignition systems>*

static electricity *n* the electrical effects caused by charges at rest *<~ causes dust to collect on a vinyl record>* □ see ELECTROSTATICS, CAPACITANCE

static memory *n* a memory which uses transistor flip-flops to store data and which does not need refresh signals □ see RANDOM-ACCESS MEMORY

station *n* a place where data enters and/or leaves a communications system *< a radio ~>* □ see WORKSTATION, EARTH STATION

stator *n* the nonmoving coil or coils of a motor, servo, or other electromechanical device – compare ROTOR

status *n* the condition of something *<the ~ of a switch>*

status register *n* FLAG REGISTER

STD – see SUBSCRIBER TRUNK DIALLING

step *vb* to execute a program a single instruction at a time – **step** *n*

stepping motor *n* an electric motor whose shaft rotates one step at a time. The number of steps per revolution (eg 48) is fixed by the manufacturer. A stepping motor contains a number of coils which are fed in a particular sequence with digital signals generated by a computer program. Stepping motors are widely used for providing precise movement of machinery (eg robot arms) under microcomputer control. □ see ROBOT ARM, NUMERICAL CONTROL, ROBOTICS

stereo *adj* of or relating to audio systems that give an impression to the listener of realistic three-dimensional sound *<~ recording>* –

stereo *n*

stochastic *adj* of or being a process or model (eg a computer simulation of radioactive decay) controlled by the laws of chance – compare DETERMINISTIC □ see SIMULATION

stop bit *n* a bit (binary 1 or 0) which is transmitted in an asynchronous communications system to tell a receiver device that a piece of information has been transmitted – compare START BIT □ see UART

storage *n* any method for holding data until it is required <~ *on magnetic disk*> □ see RANDOM-ACCESS MEMORY, READ-ONLY MEMORY, MAGNETIC DISK, MAGNETIC TAPE

storage capacity *n* the amount of computer data (usu in byte-length words) which can be stored in a computer memory device <*a ~ of 5 megabytes*> □ see MEGABYTE, MASS STORAGE

store *vb* to put data into a computer memory device for future use <*to ~ a program*> – **store** *n* □ see MEMORY

strain gauge *n* a sensor attached to an object to detect how much distortion it undergoes when stressed. A strain gauge usu consists of a fine grid of metal foil, the resistance of which changes under tension or compression. The change of resistance is measured electronically and used to help engineers investigate distortion of materials and structures (eg bridges) affected by forces. □ see TRANSDUCER

stream *n* – see FLOW, TRAIN

string *n* **1** a series of events or items <*a ~ of bytes*> **2** a set of characters stored in groups of various lengths in a series of memory locations

stringy floppy *n* a data storage device consisting of a continuous loop of magnetic tape in a small cartridge. Programs on a stringy floppy can be loaded into a computer almost as fast as from a floppy disk.

stripboard *n* a sheet of electrically insulating material, perforated with holes, and with copper tracks on one side. Circuits are assembled using stripboard by poking the wires of components through the holes and soldering them to the copper tracks. □ see PRINTED CIRCUIT BOARD

¹strobe *n* STROBOSCOPE

²strobe *vb* to read information from something <*to ~ a latch*> – **strobe** *n*

stroboscope *also* **strobe** *n* an instrument or device for measuring the speed of rotation of something. One type of stroboscope produces a series of brief bright flashes of light, the frequency of which can be adjusted electronically to make the rotating object appear to stand still, hence enabling the speed to be found. Another type of stroboscope consists of a regular light and dark pattern which is used on a record turntable to ensure that the turntable rotates at the right speed.

¹structure *vb* to select and combine parts to make something complete *<to ~ a program>* □ see STRUCTURED PROGRAMMING
²structure *n* an arrangement of parts which make a whole *<the ~ of the electronics industry>*
structured programming *n* the art of writing computer programs that are well laid out and easy to follow. One aim of structured programming is to divide the problem into self-contained units that can be written and tested separately. In this way the main program would consist of a set of instructions which sends the computer to a particular unit which is carried out before returning to the main program for a fresh instruction. □ see SUBROUTINE
stylus *n* **1** a tiny piece of a crystal (eg diamond) on a hi-fi cartridge which follows the groove on a vinyl disk **2** a penlike device used to transfer data from a digitizer to a computer **3** NEEDLE
subassembly *n* a unit built and usu working but intended as part of a more complex unit *<a liquid crystal display is a ~ of a multimeter>* □ see SYSTEM, BUILDING BLOCK
submarine cable *n* a cable submerged beneath the sea used, esp previously, for long- distance communication. Submarine cables are still being laid down despite the rapid growth in satellite communications. Indeed, the first undersea fibre optics cable was laid between Britain and Europe in 1984. □ see COMMUNICATIONS SATELLITE, OPTICAL COMMUNICATIONS
sub-micron chip *n* a silicon chip on which some components have details smaller than one millionth of a metre (a micron) wide. This extremely fine detail is possible using the technique of X-ray lithography. □ see MICRON, X-RAY LITHOGRAPHY, MOORE'S LAW
subroutine *n* a self-contained set of instructions in a computer program which can be called up and run from any part of the main program. A subroutine is usu a frequently performed task. □ see PROCEDURE
subscriber *n* a person or an organization that pays to make use of a communications system *<a telephone ~>*
subscriber trunk dialling *n* (abbr **STD**) a telephone service enabling subscribers to make long distance calls without going through an operator
subscript *n* a (smaller) letter or number printed just below another letter or number (eg '2' is a subscript in the chemical formula for silicon dioxide – SiO_2) □ see SUPERSCRIPT, EXPONENT
substrate *n* a layer of semiconductor in an integrated circuit on which transistors and other devices are formed *<the ~ of a silicon chip>* □ see PHOTOLITHOGRAPHY, EPITAXIAL LAYER, EPITAXY
subsystem *n* a part of a system which is usu capable of operating by itself *<an amplifier is a ~ of a radio>* □ see SYSTEM
suite *n* a group of items that together make a coordinated package *<a ~ of educational programs>*
summing amplifier *n* an amplifier, based on one or more integrated

circuit operational amplifiers, designed to add together a number of
voltages. Once used extensively in analogue computers, summing
amplifiers are now used mainly as mixers in audio applications for
combining the outputs from microphones, electric guitars, etc. □ see
SUMMING POINT

summing point *n* the point at the inverting input of an operational
amplifier which the op amp holds at zero volts, thereby enabling
voltages to be added together for analogue computing and audio
applications □ see SUMMING AMPLIFIER

superconducting magnet *n* an electromagnet which is operated at
close to absolute zero (ie −273C) so that the coil of wire carrying the
current which generates the magnetic field loses all its electrical
resistance. The intense magnetic field produced by a
superconducting magnet is used for research into nuclear fusion, and
in some types of diagnostic medical instrument.

superconductor *n* a metal (eg tin) which loses all its electrical
resistance at temperatures close to absolute zero (ie −273C) –
superconductive *adj*, **superconductivity** *n* □ see JOSEPHSON EFFECT,
SUPERCONDUCTING MAGNET, CRYOGENIC

super large scale integration *n* (abbr **SLSI**) the process or
technology of making devices which have more than 100 thousand
gates per silicon chip □ see MOORE'S LAW

superscript *n* a (smaller) letter or number printed just above
another letter or number (eg 'n' is a superscript in the raising of 2 to
its power in 2^n) □ see SUBSCRIPT, EXPONENT

support software *n* programs provided by a computer
manufacturer to help users develop their own applications programs
(eg for operating a numerically controlled lathe) □ see APPLICATIONS
SOFTWARE

suppressor *n* a device for reducing electrical interference in a
circuit. A suppressor may be fitted to the source of the noise (eg to a
car's ignition system), and/or to the device (eg a microcomputer)
which is to be protected from the effects of the noise. □ see NOISE,
GLITCH

surface wave *n* GROUND WAVE

sustain *n* – see ADSR

swarm *n* lots of bugs □ see BUG

sweep generator *n* an item of test gear which automatically
produces frequencies sweeping from low to high frequency, and
which is used for testing audio equipment, esp amplifiers □ see
WAVEFORM GENERATOR

¹switch *n* **1** a mechanical device for making or breaking an electrical
circuit <*a reed* ~> **2** a transistor which can be driven on or off <*a
solid-state* ~> **3** a logic gate (eg a NAND gate) which has a high or
low logic output □ see REED SWITCH, RELAY, OPTOSWITCH

²switch *vb* to change the electrical state of something <*to* ~ *off a
central heating system*> – **switchable** *adj*

switch debouncer *n* DEBOUNCER

symbol *n* **1** a simplified shape for indicating the function of a building block in a circuit or system <*a resistor* ~> **2** a letter representing a particular quantity or unit <Ω *is the* ~ *for ohm*>

symbolic *adj* relating to the use of symbols to represent items <*a* ~ *address*> □ see SYMBOL

symbolic logic *n* the use of symbols to solve nonnumerical problems □ see SYMBOL

symmetrical *also* **symmetric** *adj*, *of a waveform* having equal high and low times <*a square wave has a* ~ *waveform*> – compare ASYMMETRICAL – **symmetrically** *adv*, **symmetry** *n*

synchronization pulses *n* any electrical pulses that enable two systems or parts of systems to communicate with each other <~ *between a TV camera and a TV receiver*> □ see TELEVISION

synchronous *adj* of or being a communications system in which data is sent within fixed time intervals – compare ASYNCHRONOUS

synchronous computer *n* a computer in which the transfer and manipulation of data takes place in sequence under the control of clock pulses. All digital computers are of this type

synchronous counter *n* a binary counter used in digital counting circuits (eg clocks) that consists of a set of flip-flops which change state all at once and which are therefore faster in operation than ripple counters – compare RIPPLE COUNTER □ see BINARY COUNTER

syntax *n* the rules which determine how words, phrases, and punctuation marks are used in a high-level computer language. Any microcomputer which uses Basic provides on-screen notes and error codes if a programmer breaks these rules <*a* ~ *error*>.

synthesize *vb* to bring together several parts to make a whole <*to* ~ *music from digital signals*> – **synthesis** *n* □ see SYNTHESIZER, SPEECH SYNTHESIS

synthesizer *also* **music synthesizer** *n* a very versatile electronic instrument which can be programmed, usu through a keyboard, to mimic the sounds of most nonelectronic instruments, and to create novel musical sounds for effect or entertainment □ see ELECTRONIC MUSIC

system *n* all the parts making up a working device <*a television* ~> □ see ELECTRONIC SYSTEM

system program *n* any program which a computer uses for carrying out housekeeping or supervisory tasks; a program other than an applications program □ see DISK OPERATING SYSTEM

systems analyst *n* someone who looks into the function of something (eg an office), identifies requirements, and devises a computer system that will meet those requirements □ see ELECTRONIC OFFICE

systems variables *n* the variables, held within a particular area of a computer's memory, which represent constantly varying information (eg the position of the screen cursor)

System X *n* a computer-based telephone system being developed in Britain to replace the present electromechanical system. In System X all calls are transmitted in digital form rather than in analogue form as at present. At the exchange, computers route the calls to a destination using solid-state logic circuits rather than electromechanical relays. System X is a modular telephone system which can be extended easily, and interfaced to optical fibre communications links if necessary.

T

T the symbol for the prefix tera meaning one million million million times □ see TERABIT MEMORY

tabulation *n* the process of making a table or a list of data in one or more columns <~ *of readings on a VDU*> – **tabulate** *vb* □ see LISTING

tachometer *n* an electronic instrument for measuring the speed of rotation of something (eg the speed of a flywheel when tuning a car). The speed is usu measured in units of revolutions per minute (RPM). Most tachometers use a photodiode to sense the reflection of light off a marker so that there is no direct contact with the rotating object. □ see OPTOSWITCH

tag *n* **1** a small metal terminal fitted to a device to enable a wire to be connected to it <*a solder* ~> **2** a label which identifies an item <*a* ~ *on a disk file*> – **tag** *vb* □ see TAG STRIP

tag strip *n* a set of metal terminals to which devices can be soldered to form a circuit □ see TERMINAL STRIP

take off *n* an electrical point on a circuit or device to which electrical connection can be made <*power* ~> – **take off** *vb*

tantalum capacitor *n* an electrolytic capacitor favoured by circuit designers because of its small size and low leakage current compared with the more commonly-used aluminium electrolytic capacitors □ see ELECTROLYTIC CAPACITOR

tap *n* a terminal on the secondary winding of a transformer which selects a particular alternating current voltage <*a 12V* ~> – **tap** *vb*

¹tape *n* a long strip of material (eg plastic with a magnetizable surface or paper) which is used to record information <*a cassette* ~> □ see MAGNETIC TAPE, STRINGY FLOPPY, PAPER TAPE, FERRITE, VIDEO TAPE

²tape *vb* to record computer data, speech, radio and TV programmes, or other information for future use on tape

tape cartridge *n* – see CARTRIDGE

tape counter *n* a device on a tape recorder for keeping a record of where a program or other information stored on the tape begins and ends

tape drive *also* **tape transport** *n* a mechanism for moving a magnetic tape past a read/write head so that data can be transferred to and from the tape □ see MAGNETIC TAPE

tape recorder *n* a recording and playback device that uses magnetic tape to store audio information or computer data – compare DISK DRIVE

tape transport *n* TAPE DRIVE

task *n* a single unit of work (eg a routine) to be carried out by a computer □ see MULTITASKING

TDM – see TIME DIVISION MULTIPLEXING

technology transfer *n* ideas and inventions in one area applied to the solution of problems and the creation of new products in an unrelated area. Microelectronics is a unique example of how a product, the integrated circuit, which was stimulated by military interests, has been successfully applied in the areas of commerce, entertainment, leisure, education, research, space exploration, etc

telebanking *n* the transfer of money between accounts by means of the telephone and/or television. Some Viewdata services (eg Prestel) offer telebanking so that goods can be purchased and payed for using a modem which communicates with a central computer holding details of a person's account. □ see PRESTEL, VIEWDATA, MODEM, SMART CARD, PIN

telechirics *n* the use of remote-control devices to carry out dangerous work without risk to the operator. Telechirics includes the use of the manipulator arm on the space shuttle to 'capture' satellites for repair, the use of a remotely-controlled vehicle equipped with a TV camera to examine a suspicious object, and submersibles for examining undersea equipment (eg oil pipelines and drilling rig structures). □ see REMOTE CONTROL, ROBOT

telecommunications *n* the use of electronic and other equipment to send information through wires, the air, and interplanetary space □ see TELEPHONE, INFORMATION TECHNOLOGY, DIRECT BROADCAST SATELLITE, COMMUNICATIONS SATELLITE, OPTICAL COMMUNICATIONS

telecommunications link *n* a path established for sending information by radio or other electronic means <*a satellite* ~> □ see EARTH STATION

telecomputing *n* the sending and receiving of information (eg games programs) between computers via the telephone line or TV □ see TELESOFTWARE, TELETEXT, VIDEOTEXT, MICRONET 800

teleconferencing *n* also **videoconferencing** the use of telecommunications, esp telephones or communications satellites, to set up an 'electronic meeting' between people who do not have enough time and/or money, to travel to a face-to-face meeting. Audio-teleconferencing which involves the transmission of simple graphics, is more widely used than video-teleconferencing which provides pictures of the participants at the teleconference. □ see TELEMEDICINE, COMMUNICATIONS SATELLITE

telegraphy *n* any system for the two-way transmission of messages (usu in printed form) between two or more users – compare TELEPHONY □ see FACSIMILE, TELEX

telematics *n* the social and economic importance of computers and telecommunications systems. Telematics is derived from the French word 'télématique' which has much the same meaning as 'information technology'. □ see INFORMATION TECHNOLOGY

telemedicine *n* the use of telecommunications to send medical information to doctors, consultants, and surgeons. For example, in the USA, eye specialists in different parts of the country can use a satellite communications system to look in on delicate eye operations carried out at the University of Missouri. □ see TELECONFERENCING

telemetering *n* the use of communications systems to send the results of measurements to a centre for recording and (computer) analysis □ see TELEMETRY, SEISMIC SURVEY

telemetry *n* the transmission by radio of measurements made at a distance. An automatic weather station sends its measurements of wind speed, temperature, air pressure, etc using telemetry. Similarly, telemetry makes it possible to maintain two-way contact with orbiting space stations (eg the shuttle) and interplanetary space probes (eg Voyager). □ see COMMUNICATIONS SATELLITE, WEATHER SATELLITE, VOYAGER, SPACE SHUTTLE

telephone *n* **1** the handset used to communicate with people or computers over the telephone system **2** the system for transmitting voice signals in the audio frequency range by wire and/or radio *<a radio ~>* □ see SYSTEM X

telephone exchange *n* a place where telephone lines and switching facilities are brought together to provide a telephone service for a particular area or city □ see SYSTEM X

telephony *n* any system for the two-way transmission of sound messages between two or more users – compare TELEGRAPHY □ see TELEPHONE

teleprinter *n* TELETYPEWRITER

teleshopping *n* a two-way computer link between the home or office and a warehouse which enables goods to be ordered via the warehouse stock control computer □ see PRESTEL

telesoftware *n* the transmission of computer software from one computer to another by telephone or television □ see MICRONET 800, PRESTEL, CEEFAX, ORACLE, MODEM

teletext *n* also **broadcast Videotext** an information service which uses a modified TV set to receive information which is carried 'piggy back' style on the normal picture information. Teletext is noninteractive since the user can access only a limited number of pages of information which can be called up using a push-button call-pad. Ceefax and Oracle are examples of Teletext – compare VIDEOTEXT

teletext decoder *n* a device connected to a television so that it can receive teletext (eg Ceefax and Oracle) □ see TELETEXT

teletypewriter *n* also **teleprinter** a device consisting of a printer and keyboard for transmitting and receiving printed information over a telephone network □ see TELEX

television *n* (abbr **TV**) the transmission and reception of pictures and sound by very high frequency or ultra high frequency radio carrier waves. These carrier waves are in bands (eg radio frequencies in the range 470 to 610 megahertz). In the British 625 line system, each broadcast channel is allocated an 8 megahertz bandwidth. The

video (picture) carrier wave is separated by 6 megahertz from the
sound carrier wave. The sound carrier wave has a bandwidth from 20
hertz to 15 kilohertz, and the video carrier wave has a bandwidth of
5.5 megahertz. The video signal is amplitude modulated while the
sound signal is usu frequency modulated. Since television was first
introduced in the 1930s it has exploited all the developments in
microelectronics, and it looks as if the microprocessor is also poised
to influence television receiver design. □ see CARRIER WAVE,
BANDWIDTH, TELEVISION RECEIVER, VIDEO
television channel *n* a path between a television camera and a
VDU which carries sound and vision signals as modulated radio
waves □ see TELEVISION, MODULATE
television receiver *n* a device for receiving pictures and sound
through the air and through cables by radio. A television receiver
has a tuned circuit consisting of a filter and amplifier to select
programme channels. A demodulator separates the sound and
picture information from the carrier wave. The sound information is
fed to an audio amplifier to operate a loudspeaker, and the picture
(video) information, along with synchronizing pulses, controls an
electron beam which writes pictures on the screen of a cathode-ray
tube. The electron beam builds up a picture by moving rapidly back
and forth (scanning) in a series of closely spaced lines. In the 625 line
system, 625 lines make up a complete picture (frame) every 1/25th of
a second. The synchronizing pulses ensure that a particular line starts
and ends at the same time as in the television camera. Whilst a
monochrome (black and white) television receiver simply produces
shades of grey by conrolling the intensity of the electron beam
reaching the phosphor on the screen, a colour television receiver also
has to produce different colours. □ see TELEVISION, COLOUR
TELEVISION, CARRIER WAVE, BANDWIDTH, ELECTRON GUN, VIDEO,
MONITOR, VDU
televize *vb* to broadcast by television □ see VIDEO
telex *n* [*tele*typewriter *ex*change service] an international system for
sending and receiving information over the telephone system using
teletypewriters □ see TELETYPEWRITER
temperature coefficient *n* the percentage change with temperature
of a particular characteristic (eg resistance) of a device or circuit. For
some applications (eg temperature measurement), a device (eg a
thermistor) is chosen which has a high temperature coefficient, but
for most electronic circuits (eg frequency generators) the
temperature coefficient is kept to the lowest possible value by careful
circuit design.
terabit memory *n* a data storage device capable of holding a billion
(10^{12}) bits of data
¹terminal *adj* being at the end of something <*a ~ connector*>
²terminal *n* **1** a device for communicating with a computer. A
terminal can be a large desktop unit consisting of a full keyboard and

VDU, or it can be a small handheld device consisting of a keypad and digital display which can be linked to a host computer via a modem connected to the telephone line *<a stock control ~>*. □ see GRAPHICS TERMINAL **2** a metal wire or pin on a device enabling it to be connected to another device fitted with an appropriate connector □ see TERMINAL STRIP

terminal strip *n* a set of screws mounted on a length of electrically insulating material to which devices can be connected to make up circuits □ see TAG STRIP

terminate *vb* **1** to stop something happening *<to ~ a program>* **2** to connect resistors to the end of a wire or cable so as to adjust the impedance of the wire to a predetermined value *<to ~ an aerial lead>* – **termination** *n*

test *vb* to find out whether a circuit or program is working properly – **test** *n*

test equipment *n* any apparatus (eg a multimeter) designed for faultfinding in circuits □ see MULTIMETER, CATHODE-RAY OSCILLOSCOPE, WAVEFORM GENERATOR

text *n* **1** a self-contained message printed or displayed in alphanumeric form **2** information stored in memory ready for printing or display □ see TEXT EDITOR, WORDPROCESSOR

text editor *n* a program in a text-handling system (eg a wordprocessor) which carries out editing functions (eg deleting words, moving text in blocks, and justifying lines)

text processing *n* – see WORDPROCESSING

thermal *adj* of or caused by heat or (high) temperature *<~ noise>*

thermal matrix printer *n* a printer which forms characters by burning their shapes on heat- sensitive paper as a pattern of small dots □ see DOT MATRIX PRINTER

thermal noise *n* electrical interference on a communications channel caused by the agitation of molecules in components □ see WHITE NOISE

thermal paper *n* also **thermographic paper** heat-sensitive paper which is chemically treated so that it darkens when heat is applied □ see THERMAL MATRIX PRINTER

thermal runaway *n* the destruction of a bipolar transistor caused by the progressive rise in current through it as its temperature increases. The normal working current flowing through a transistor causes it to warm up slightly which releases more electrons and holes for conduction. The resistance of the transistor decreases causing more current to flow and hence more heat to be produced. The process can make the transistor burn out unless care is taken to counteract the increasing current through careful circuit design. Silicon transistors are not as liable to thermal runaway as the older germanium transistors. □ see HEAT SINK

thermionic emission *n* the release of electrons from a heated metal. Valves and cathode-ray tubes have a heated metal filament from which a stream of electrons is focussed and controlled by electrodes. □ see TRIODE VALUE, CATHODE-RAY TUBE

thermistor *n* a heat sensor made from a mixture of semiconductors and having an electrical resistance which varies with temperature. Thermistors are used for temperature measurement (thermometers) and control (thermostats), and for protecting equipment from sudden current and voltage surges – compare THERMOCOUPLE □ see THERMOSTAT, TEMPERATURE COEFFICIENT

thermocouple *n* a device made from a pair of dissimilar metals (eg copper and iron) which produces a voltage varying with temperature. The metals are joined end to end to make two junctions. Any difference of temperature between the junctions produces a voltage which varies according to the temperature difference.
Thermocouples are used in electronic thermometers and have the advantage compared with thermistors that they can measure higher temperature and that the heat-sensitive junction is very small. □ see THERMISTOR

thermoelectric refrigerator *n* a usu small device that consists of two different metals or semiconductors joined end-to-end and through which current is made to flow to cool one of the junctions between the metals. The cooling effect can be usefully used to refrigerate any electronic equipment in contact with this cold junction; the heat removed at the cold junction is released at the hot junction. □ see THERMOCOUPLE

thermographic paper *n* THERMAL PAPER

thermostat *n* an electronic device for controlling temperature <*a fish tank* ~> □ see THERMISTOR

thick-film resistor *n* an electrical resistor made by depositing a film of metal (eg silicon-chromium) on an insulating base. The metal is deposited from its vapour in a vacuum, and the correct resistance is obtained by adjusting the thickness of the film. Thick-film resistors are available singly or in packages (eg a single-in-line package) to make the assembly of complex circuits easier. □ see THIN-FILM RESISTOR

thin-film resistor *n* an electrical resistor formed by depositing a thin film of resistive material (eg silicon-chromium) on an insulating base. The resistance required is obtained by using a thin beam of laser light to burn away the film to change its width and length. This technique is used to make single resistors and for forming resistors on a silicon chip in the making of integrated circuits. □ see THICK-FILM RESISTOR

third generation *adj* – see COMPUTER GENERATIONS

three-D television *n* any method of making TV pictures appear to have depth and perspective. Most of the techniques which have been developed for three-D television depend on the viewer wearing special glasses so that each eye is presented with a slightly different viewpoint of a scene (eg one view in green light and the other in red light). The eyes bring these two images together to give an impression of depth. However, holography offers the most exciting prospect for three-D television for it will make it possible to project 3D images into a room, round which we shall even be able to walk. □ see HOLOGRAM, TELEVISION

three-state *adj* TRISTATE

threshold *n* a particular value of signal strength which causes a response in a circuit <*the lower ~ of a logic gate*> □ see SCHMITT TRIGGER, COMPARATOR

throughput *n* **1** the amount of work a computer gets through per unit time **2** the amount of data a communications system transmits per unit time □ see INPUT, OUTPUT

thyristor *n* SILICON-CONTROLLED RECTIFIER

time base *n* a regular series of pulses for **a** controlling the movement of an electron beam across the screen of a cathode-ray tube <*a ~ generator*> □ see RASTER SCANNING, CATHODE-RAY TUBE **b** coordinating the operation of shift registers in computers and other digital devices □ see CRYSTAL CLOCK, SHIFT REGISTER

time bomb *n* an ingenious method for protecting business software against piracy, ie unauthorized copying, that consists of a piece of code within a package which would normally be removed when installed by a bona fide dealer. On a pirated copy however, the time bomb, after a certain period during which the company will have become very dependent on the use of the package, will 'explode', turning the company's files into meaningless garbage and probably ruining the disks holding the programs as well. □ see DONGLE

time constant *n* the time taken for the voltage across a capacitor to rise to 63 per cent of its final voltage when it charges through a resistor connected in series with it. Time constant is given by the simple equation: time constant = value of resistor × value of capacitor. A knowledge of time constants is useful in the design of timing circuits (eg clocks). □ see TIMER, CAPACITOR

time delay circuit *n* – see MONOSTABLE

time division multiplexing *n* (abbr **TDM**) a method of sending more than one message along a communications channel that involves interleaving the digital pulses of each message by giving each message a time slot every few microseconds. At the receiving end, the time slots for each train of pulses are separated out to recreate the original messages – compare FREQUENCY DIVISION MULTIPLEXING

time out *vb* to complete a time delay <*the monostable* timed out> □ see TIMER

timer *n* a circuit or device which provides a single time delay or a regular series of time delays for controlling a sequence of events. Timers in integrated circuit form are widely used in digital clocks and watches, computers, counters, washing machines, robots, and industrial controllers. Most timers make use of a capacitor for temporarily storing a charge. □ see CRYSTAL CLOCK, TIME BASE, MONOSTABLE, ASTABLE

timesharing *n* the use of a computer by a number of people at (apparently) the same time. Because a computer can process data at very high speed, it is possible to make it attend to the needs of a number of users in rapid sequence, although it appears to any one user that he/she has sole use of it. □ see MULTITASKING, ACCESS TIME, TIME SLICE

time slice *n* the amount of time allocated to each user when a number of users share a computer's facilities □ see TIMESHARING, TIME SLOT

time slot *n* a brief interval of time during which a device is able to accept or deliver digital signals □ see TIME DIVISION MULTIPLEXING

time switch *n* an electronic device for controlling the length of time that a device (eg a room light) is to remain on and off

timing pulse *n* a brief and usu repetitive signal used for timing operations in electronics circuits. Timing pulses are produced by a crystal clock in a microcomputer. □ see CRYSTAL CLOCK

tin *vb* to coat the tip of a soldering iron, a copper track on a circuit board, or the end of a wire with a thin layer of solder to help prepare the surfaces for soldering – **tinning** *n* □ see SOLDER

¹toggle *n* FLIP-FLOP

²toggle *vb* to change from one state to another □ see FLIP-FLOP

toggle switch *n* a switch with a small projecting arm or knob which is moved 'up', 'down', and possibly 'centre' to control current flow in a circuit □ see SWITCH

tomography *n* – see COMPUTERIZED AXIAL TOMOGRAPHY

tone control *n* a control device on audio amplifiers, esp a hi-fi amplifier, for adjusting the relative strengths of the high (treble) and low (bass) audio frequencies

tool kit *n* a software package which enhances a microcomputer's facilities. Tool kits consist of machine code routines (eg for debugging), or graphics (eg sprites), or sound capabilities (eg 'whizz-bang' sounds for arcade type games).

total internal reflection *n* the complete reflection of a beam of light from inside a transparent material when the light strikes the surface of the material at an angle which is greater than a certain (critical) angle. The transmission of laser signals down an optical fibre, without leakage from the sides of the fibre, in optical communications systems is dependent on total internal reflection. □ see OPTICAL COMMUNICATIONS, OPTICAL FIBRE

totem pole *n* a way of connecting two bipolar transistors, one 'on top of the other', in the output stage of a TTL logic gate so as to speed up its operation when it is driving capacitive loads □ see GATE, TRANSISTOR-TRANSISTOR LOGIC

touch-screen computer *n* a (personal) computer which, though having a conventional keyboard, can be operated by touching images (icons) on its VDU which represent a menu of options □ see ICON, MOUSE, WINDOW

touch-sensitive *adj, of keyboards, keypads and switches* requiring only light pressure to operate them □ see MEMBRANE KEYPAD

TPI [*t*racks *p*er *i*nch] – see TRACK DENSITY

trace *n* the line left by an electron beam moving across the screen of a cathode-ray tube or other recording device □ see RADAR, PHOSPHOR, CATHODE-RAY TUBE

¹track *n* **1** a narrow path on the surface of a magnetic disk, tape, compact disk, vinyl disk, or video disk which carries, or can carry, information □ see MAGNETIC DISK, MAGNETIC TAPE, COMPACT DISK, VIDEO DISK, DIGITAL RECORDING **2** the path followed by something (eg an interplanetary probe) **3** a flexible belt on a vehicle (eg a robot) on which it travels

²track *vb* to follow the path of an object, usu by radio <*to* ~ *a migrating animal*>

track ball roller *n* a hand-operated device used in computer games to move a graphics shape round the screen of a VDU. The device has a sphere about the size of a billiard ball which is rolled in the palm of the hand in the direction the shape is to be moved on the screen. The ball drives two wheels set at right angles to each other and connected to potentiometers. The signals produced are fed to the microcomputer to control the sideways and upwards movement of the shape. Buttons are usu provided for firing at a target in a game. The track ball roller provides finer, faster, and more accurate positioning of the shape compared with a joystick – compare JOYSTICK □ see POTENTIOMETER

track density *n* the number of tracks per inch (TPI) on a magnetic disk measured along a radius. For floppy disks, the track density is usu 48, while for hard disks it varies from 100 to 400. □ see MAGNETIC DISK, TRACK, TRACK WIDTH

tracking *n* the process of ensuring that a sensing device (eg a read/write head) keeps track of information it is supposed to follow

track width *n* the width of a single track on a magnetic disk. The track width of floppy disks is about 0.013 inches. For hard disks, the track width is about 0.004 inches, while for Winchester disks it is as low as 0.0018 inches. □ see TRACK DENSITY

tractor feed *n* also **sprocket feed** a method of moving paper through a printer using pins on the platen to engage sprocket holes at the edge of the paper

traffic *n* the use made of a communications system, usu measured in terms of the number of simultaneous telephone conversations the system can handle □ see COMMUNICATIONS SATELLITE

trailing edge *n* **1** the last part of a signal waveform before its amplitude drops to zero **2** the edge of a punched card which is the last to enter a card reader – compare LEADING EDGE

train *n* also **stream** a sequence of events following one after the other <*a pulse* ~>

transceiver *n* **1** a telephone or two-way radio which can both send and receive messages **2** a computer circuit which passes data to and from memory

transconductance *n* a measure of the performance of a field-effect transistor, defined as the change in the drain current brought about by a change in the gate-to-source voltage □ see FIELD-EFFECT TRANSISTOR

transducer *n* a device for changing one form of energy into another. Transducers are used extensively in electronic instrumentation and control and communications systems. For example, a thermocouple is used in electronic thermometers since it changes heat energy into electricity; a solar cell is used for automatic exposure control in cameras since it changes light energy into electricity; and a microphone is used in a telephone since it changes sound energy into electricity.

transfer *vb* **1** to copy data in one register of a microprocessor and put it into another register **2** to apply the skills and knowledge obtained in one field of science and technology to another unrelated field <*space technology* ~ed *to commercial aircraft design*> – **transfer** *n* □ see TECHNOLOGY TRANSFER

transfer characteristic *n* a graph which shows the relationship between the input and output current or voltage for a semiconductor device <*the* ~ *of an op amp shows the relationship between input and output voltages*>

transfer rate *n* the speed (eg in bits per second) at which data is moved from one place to another □ see BAUD

transformer *n* an (electromagnetic) device for converting alternating current from one voltage to another. A transformer consists of two coils wound on an iron core. One coil, the primary, carries the input voltage, and the other coil, the secondary, provides the output voltage. The input and output voltages are in the same ratio as the ratio of the number of turns of wire on the primary and secondary coils. Thus a transformer can be used to step up (ie increase) or step down (ie decrease) the voltage of an alternating current. Transformers are used in power supplies and audio systems. □ see ELECTROMAGNETIC INDUCTION

transient *n* an unwanted short-lived electrical signal in a circuit caused by faulty components or electrical interference from nearby circuits. Sensitive equipment such as computers need protection against the effects of transients. – **transient** *adj* □ see INTERFERENCE, SUPPRESSOR, GLITCH

transient state *n* a short time interval during which a circuit or device changes from one operating mode to another <*the* ~ *of a flip-flop*> □ see TRANSITION

transistor *n* a semiconductor device which has three terminals and works as a switch or amplifier □ see BIPOLAR TRANSISTOR, FIELD-

EFFECT TRANSISTOR

transistorized *adj* operated by transistors, esp single transistors rather than integrated circuits which contain transistors <~ *electronics*>

transistor-transistor logic *n* (abbr **TTL** or **T²L**) a common type of logic gate in integrated circuit form that is characterized by high speed, low power dissipation, and a nominal 5 volt power supply. The basic building block of transistor-transistor logic consists of two bipolar transistors, one of which is a multi-emitter device which forms the inputs to the gate (eg a 4-input NAND gate). Digital circuits designed from TTL are used in a wide variety of devices from simple burglar alarms to industrial robots – compare COMPLEMENTARY METAL-OXIDE SEMICONDUCTOR LOGIC □ see GATE 1, EMITTER-COUPLED LOGIC, SCHOTTKY DIODE

transition *n* a change from one condition to another <*energy level* ~s *in atoms*> – **transitional** *adj*

transition frequency *n* the frequency at which the alternating current gain of a bipolar transistor fails to unity. Manufacturers of transistors publish the transition frequency of their devices to help circuit designers choose devices for operation at high frequency.

translate *vb* **1** to change one programming language into another, esp source code into object code □ see COMPILER, ASSEMBLER **2** to move a part of a graphics image on a VDU to another part of the screen – **translation** *n*, **translator** *n*

transmission *n* the sending of information from one place to another – **transmit** *vb* □ see TRANSMITTER, COMMUNICATIONS SYSTEM

transmission line *n* □ see LINE, LINK

transmission speed *n* the number of items (eg words) sent along a communications link per second □ see BAUD, PROPAGATION SPEED

transmitter *n* **1** that part of a communications system which sends out messages <*a TV* ~> **2** an integrated circuit designed to transmit signals from a sensor along just two wires which also carry the power supply to the sensor

transmitting station *n* **1** a radio transmitter for sending out radio messages **2** any device (eg a computer terminal) which sends data to receivers in a communications network

transparent *adj* relating to the operation of a computer without the user being aware of intervening software or hardware <*a computer's operating system is* ~ *in operation*> – **transparency** *n*

transponder *n* a combined receiver and transmitter which transmits signals automatically when triggered. Transponders are used in communications satellites for relaying messages between Earth stations and the satellites. □ see DIRECT BROADCAST SATELLITE

transport *n* – see TAPE DRIVE

transportable *adj* able to be carried <*a* ~ *data logger*> □ see PORTABLE, PORTABLE MICRO

transputer *n* a powerful microprocessor being developed in Britain for the fifth generation computers of the 1990s. Its name is derived from 'transistor' and 'computer' which underlines the idea of a revolutionary microprocessor which will become a universal building block used with others in large quantities just as transistors are today. Transputer actually refers to a family of products, the first of which is to be a 32-bit microprocessor containing a quarter of a million transistors integrated on a chip 45 square millimetres in area and capable of executing 10 million instructions per second. The transputer is not designed to be the prima donna of a microcomputer system, sitting in the middle of a large supporting cast of memory and peripheral devices, although initial applications will use it like this. Instead, special communications circuits are included in the design so that it can work as part of a team of many other transputers all working closely together to increase processing power. Such an array of transputers will enable supercomputers to be built that are capable of allowing intelligent interaction between people and machines. □ see FIFTH GENERATION COMPUTER, VERY LARGE SCALE INTEGRATION

triac *n* a three-terminal semiconductor device used for controlling alternating current power <*a ~ controlled electric drill*> □ see SILICON-CONTROLLED RECTIFIER

trichloroethylene *n* – see TRIKE

¹trigger *adj* of or being a signal that sets off some activity in a circuit <*a ~ pulse*>

²trigger *vb* to set off activity in a circuit with a signal from another circuit <*to ~ a monostable*>

³trigger *n* **1** an input terminal on a flip-flop which is used to make the flip-flop toggle, ie to change state □ see FLIP-FLOP **2** TRIGGER CIRCUIT

trigger circuit *also* **trigger** *n* a circuit (esp a flip-flop) which changes from one stable state to another on receiving a single trigger pulse □ see TRIGGER

triggering level *n* the value of the voltage required to make a flip-flop change state □ see EDGE TRIGGERING

trike *n* [trichloroethylene] a chemical used to 'develop' selected areas of photoresist on the surface of a silicon chip after exposure to ultraviolet light. The ultraviolet light passes through a photomask to react with a layer of photoresist. After treatment with trike, an acid is used to remove the developed areas of photoresist. In the neighbourhood of factories and laboratories which manufacture integrated circuits, there is some concern that trike, and other chemicals used by the industry, cause ill-health and pollution. □ see PHOTORESIST, ETCHING, PHOTOMASK, PHOTOLITHOGRAPHY, ULTRAVIOLET

trim *vb* to make small adjustments to something in order to achieve optimum results <*~ming an oscillator*> □ see TRIMMER

trimmer *n* a small variable resistor or variable capacitor used for making fine adjustments to the operation of a circuit *<fine tuning a radio with a ~>* □ see ²VARIABLE

triode *also* **triode valve** *n* a three-terminal device in which electrons moving in a vacuum are controlled by electrodes. The triode valve was invented at the turn of the century by modifying a diode valve with a third electrode, the grid, placed between the anode and cathode electrodes. Small signals applied to the grid controlled a much larger current flowing between the cathode and anode. The triode valve played a very important part in the development of wireless, television, and computing until the transistor's rise to fame in the 1960s. □ see DIODE, THERMIONIC EMISSION, BIPOLAR TRANSISTOR, COMPUTER GENERATIONS, SILICON CHIP, DE FOREST

trip computer *n* a microprocessor-based instrument fitted to some motor vehicles which has push-buttons and a digital display to let a driver know how much fuel is being used and how long a trip will take. The microprocessor receives the driver's estimate of the distance of the journey and data from sensors fitted to the fuel pipe and road wheels. The time of day is also displayed and the trip computer might also report progress during the trip in its own synthesized voice. □ see ELECTRONIC ENGINE MANAGEMENT

tristate *adj* also **three-state** *esp of an interface device in a computer system* having three electrical states. As well as sending or receiving binary data, such a device also has a high impedance state that disconnects it from the computer bus so allowing other devices to be active in transferring data *<a ~ buffer>*. □ see INPUT/OUTPUT PORT, BUFFER

trivalent *adj, of an atom* having a combining power of three – compare PENTAVALENT □ see VALENCY, DOPING

troubleshoot *vb* to locate and correct errors in software or hardware *<~ing a printer>* – **troubleshooter** *n*

true *n* the state of a digital logic circuit represented by binary 1 – compare FALSE □ see LOGIC LEVELS

trunk line *n* a communications channel which handles many thousands of telephone calls over long distances

truth table *n* a list of 0s and 1s showing how a logic gate or a digital circuit responds to all possible combinations of binary input signals. A truth table is useful in the design of burglar alarms, washing machines, combination locks, and robots where digital circuits are used to make decisions based on a number of inputs. In these applications, the truth table is used to predict when a control signal is present, represented by binary 1 (eg a switch is 'on'), or absent, represented by binary 0 (eg a switch is 'off'). □ see GATE 1, KARNAUGH MAP

TTL – see TRANSISTOR-TRANSISTOR LOGIC

TTL compatible *adj, of complementary metal-oxide semiconductor integrated circuits* able to be operated with transistor-transistor logic

(TTL) in the same circuit using the same power supplies

T-type flip-flop a little-used variety of flip-flop (or bistable) which is designed so that its outputs toggle on each low to high change of a clock signal applied to its input. The JK flip-flop can be made to behave as a T- type flip-flop. □ see JK FLIP-FLOP

tube *n* any device which works by controlling a beam of electrons in an evacuated tube □ see CATHODE-RAY TUBE, TRIODE VALVE, THERMIONIC EMISSION

tuned circuit *n* a circuit containing an inductor and a capacitor which can be tuned to receive particular radio broadcast programs. Tuning is usu done by varying the value of the capacitor. □ see ^{2}VARIABLE, VARICAP DIODE, INDUCTOR, SELECTIVITY

tuning *n* any method of adjusting a device or system so that it operates at peak performance <~ *a radio*> – **tuner** *n,* **tune** *vb*

Turing, Alan (1912-54). An English mathematician who worked on cracking German radio messages in coded form during World War II using a machine called ENIGMA. Later, Turing was responsible for developing a more advanced code breaker, the Colossus, which was the first operational computer to be built in England using valves. During the 1950s, Turing used computers to solve problems in the field of biochemistry. But he is perhaps best remembered for his theoretical work on computers during the 1930s. □ see TURING MACHINE

Turing machine *n* a theoretical model of a computer system used for deciding if a problem is 'computable' by controlling the movement of a hypothetical tape □ see TURING, KNOWLEDGE ENGINEERING

turnkey system *n* a computer system, esp a business computer, which, once installed, requires the simple equivalent of a 'turn of a key' by an inexperienced person to get it working

turtle *n* a mechanical robot with wheels (a floor turtle), or a graphics shape on the screen of a VDU (a screen turtle), which traces a path (a turtle trail) controlled by commands from a computer. The concept of a turtle was developed in 1960s by Seymour Papert who believed that geometric ideas are far more comprehensible to a child if they are based on body motions which could be experienced by the child. To help him achieve his aims, Papert developed the high-level language Logo as a bridge between abstract mathematics and the material world of the five senses. Children learn geometry for example, by telling a turtle what to do using simple commands from the computer. □ see PAPERT, LOGO

TV – see TELEVISION

TV cable system *n* a television system using coaxial cables to deliver television pictures, usu in a local area. A TV cable system is often used in areas of poor radio reception, or to provide programs of special or local interest, or in industrial applications and security systems. □ see DIRECT BROADCAST SATELLITE, COAXIAL CABLE

tweak *vb* to make fine adjustments to software or hardware □ see
TRIM

tweening *n* [in-betweening] a process (undertaken by a person
called an in-betweener) in computer graphics whereby a whole block
of (animated) graphics is fabricated on the screen by specifying the
first and last frames of the sequence

tweeter *n* a loudspeaker (smaller than a woofer) used in high
quality audio systems for radiating high frequency sound – compare
WOOFER □ see LOUDSPEAKER

twisted pair *n* a cable, usu contained in a single sheath, which
consists of two wires twisted together and which is used to reduce the
pickup of interference

type ahead buffer *n* an interface device in some computer systems
for the temporary storage of data entered from the keyboard should
the computer be busy doing something else □ see BUFFER 2

U

UART [*u*niversal *a*synchronous *r*eceiver and *t*ransmitter] a single integrated circuit used in some computer systems for providing the correct signals to operate asynchronous devices (eg printers). A UART accepts binary data in parallel form from the computer and sends it out in serial form. The serial data contains start, stop, and parity bits to enable a receiver to check that a data word has been sent before converting it to a parallel data word. □ see SERIAL SIGNALS, SERIAL INTERFACE, ASYNCHRONOUS, USART

UHF *adj or n* [*u*ltra *h*igh *f*requency] (of or being radio waves having) a frequency in the range 300 million hertz to 3000 million hertz <*television broadcasts use ~ radio waves*> □ see UHF MODULATOR, TELEVISION

UHF modulator *n* [*u*ltra *h*igh *f*requency modulator] a device inside most microcomputers enabling them to use a normal TV receiver as a VDU for displaying program listings and graphics □ see UHF, MONITOR

ULA – see UNCOMMITTED LOGIC ARRAY

ultra high frequency *adj or n* UHF

ultrasonic scanner *n* a computerized medical instrument which provides an image of internal organs of the body using high frequency sound waves. The waves are generated at frequencies between about 1 million hertz (1MHz) and ten million hertz by a special transducer held in contact with the body. The echo pattern from the ultrasonic waves is processed electronically to give an image on a VDU. The ultrasonic scanner is much safer than the use of X-rays as a diagnostic instrument, esp in the examination of a developing baby in the womb.

ultrasonic transducer *n* a device for transmitting and/or receiving sound waves which have frequencies beyond the range of human hearing. Ultrasonic transducers are usu available as matched pairs, ie one acts as the receiver which is tuned to the frequency generated by the other, the transmitter. Ultrasonic transmitters are widely used in burglar alarm systems and, less commonly nowadays, for remote control of TV sets. □ see ULTRASONIC SCANNER

ultrasonic waves *n* sound waves which have frequencies above about 20 thousand hertz and which are therefore inaudible to the human ear □ see ULTRASONIC TRANSDUCER, ULTRASONIC SCANNER

ultraviolet *adj* of or being electromagnetic radiation having wavelengths in the range between violet light and X-rays, ie between 400 nanometres and 2 nanometres, respectively. Ultraviolet rays in sunlight are responsible for a suntan. These rays are more energetic than visible light so they are used to make chemical changes to a photoresist in the manufacture of integrated circuits and printed circuit boards, and also for erasing data from some types of read-only memory – compare INFRARED □ see PHOTOLITHOGRAPHY, PHOTORESIST, EPROM

umbilical cord *n* a flexible cable which carries electrical power and/or signals to and from a device <*an ~ to an industrial robot*>

uncommitted logic array *n* (abbr **ULA**) a type of integrated circuit containing a large number (an array) of logic gates which can be connected together on the chip according to a customer's particular needs. An uncommitted logic array is committed to a particular application by using a metal mask to photoengrave a pattern on the metal layer forming the interconnections between the gates. Since it is only necessary to create a single mask for each array, a manufacturer who does not have full facilities for making integrated circuits can hold uncommitted logic arrays in stock and program the chip to perform an extremely diverse range of functions. Thus the manufacturer could produce a sophisticated computer from just four major components; a microprocessor, some ROM, some RAM and a ULA. □ see CUSTOM DESIGN, GATE 1, INTEGRATED CIRCUIT, PROGRAMMABLE LOGIC ARRAY

unconditional branch *n* a jump from one point to another in a program without regard to what the next instruction is <*the command GOTO in Basic is an ~*> – compare CONDITIONAL BRANCH

unijunction *n* a semiconductor device used for producing sawtooth waveforms. The unijunction has three terminals, an emitter connection and two base connections, and it is used in waveform generators and, sometimes, for controlling the movement of the electrons in cathode-ray tubes.

unipolar *adj, of a transistor* depending for its operation on either p-type or n-type semiconductor as in a field-effect transistor – compare BIPOLAR □ see FIELD-EFFECT TRANSISTOR

UOSAT [*U*niversity *o*f Surrey *sat*ellite] a communications satellite in orbit round the Earth which transmits information in English by means of a speech synthesizer. The readings of 59 instruments and 45 switches give data about solar radiation, the temperature on board the satellite, etc. UOSAT can be tuned in by radio amateurs anywhere in the world. □ see SPEECH SYNTHESIS, TELEMETRY, COMMUNICATIONS SATELLITE

up and running *adj, esp of a computer* switched on and doing the job for which it is designed

update *vb* to replace old information by new <*to ~ a database*> – **update** *n*

up-down counter *n* a binary counter used in digital systems (eg timers and process control circuits) which can count up (ie add to a previous count) or count down (ie subtract from a previous count) □ see BINARY COUNTER

upgrade *vb* to expand the capabilities of a microcomputer by adding extra memory or other peripheral devices – **upgrade** *n* □ see USER

PORT, PERIPHERAL

uplink *n* – see EARTH STATION

up time *n* any period of time during which a computer system is ready for use – compare DOWN TIME

USART [*u*niversal *s*ynchronous/*a*synchronous *r*eceiver and *t*ransmitter] a UART which is capable of both synchronous and asynchronous operation □ see UART

user *n* a person or group of people who make use of a particular computer or computer system's facilities □ see USER-FRIENDLY, USER-GROUP

user-defined graphics *n* the creation by a programmer of more interesting and/or higher resolution images on a VDU than can be obtained from a microcomputer's predefined graphics stored in its ROM

user-friendly *adj, of a computer, program, etc* providing information and facilities helpful to its user. For example, a user-friendly program is easy to use and enables errors to be located and corrected quickly. □ see SPEECH SYNTHESIS, FIFTH GENERATION COMPUTER

user-group *n* a group of people who use one particular computer, or similar computers, and who share their knowledge and programming skills through meetings and publications

user-oriented language *n* a computer language, esp a high-level language, which is written with the interests of the user, rather than the needs of the machine, in mind □ see HIGH-LEVEL LANGUAGE, USER-FRIENDLY, FIFTH GENERATION COMPUTER

user port *n* a connector on a computer into which add-ons (eg a joystick) can be plugged to extend the computer's capabilities. Some user ports are accessible through a slot in the case of the computer to a part (an edge connector) of the computer's printed circuit board. Other user ports consist of special connectors (eg a D-type connector). □ see PERIPHERAL, EDGE CONNECTOR

user program *n* APPLICATIONS PROGRAM

user terminal *n* a computer terminal at a user's home or workplace *<a travel agent's ~>* □ see TERMINAL

utility *n* a usu short and frequently used program for sorting data, making copies of files, or similar tasks. A utility is generally provided as an integral part of a computer system designed for business applications □ see HOUSEKEEPING, DISK OPERATING SYSTEM, ROUTINE

UV-erasable *adj, of a memory device* capable of having data erased by exposure to ultraviolet light □ see EPROM, FAMOS

V

V the symbol for the unit of electrical potential, the volt □ see VOLT

V24 a well-known recommendation by the CCITT (International Telegraph and Telephone Consultative Committee) which specifies the way asynchronous connections should be made between computers and peripheral equipment such as printers and modems. Most manufacturers use a 25-pin D-type connector to implement the connections. □ see RS-232 INTERFACE

vacuum tube *n* any electronic device (eg a cathode-ray tube) which controls the flow of electrons moving between electrodes in a vacuum □ see CATHODE-RAY TUBE, THERMIONIC EMISSION

valency *n* a number which measures the combining power of an atom and which is determined by the number and energies of an atom's outermost layer (shell) of electrons. Silicon has a valency of 4 since it has 4 electrons in its outermost layer all of which can be shared with 4 neighbouring atoms. A p-type or an n-type semiconductor, used in transistors and integrated circuits, is made by introducing a very small quantity of a certain type of 'impurity' atom into the crystal structure of silicon to change its electrical properties in a certain way. For example, if an impurity atom which has a valency of 3 (eg boron) is introduced into silicon, a p-type semiconductor is made which does not have sufficient electrons in its crystal structure for all its atoms to combine with neighbouring silicon atoms. Similarly, if the impurity atom has a valency of 5 (eg phosphorus), silicon becomes an n-type semiconductor which has an excess of free electrons in its crystal structure. □ see P-TYPE, N-TYPE, PN JUNCTION, BIPOLAR TRANSISTOR, CONDUCTION BAND

valency electrons *n* electrons which occupy the outermost energy level of an atom and which take part in the formation of bonds between atoms □ see VALENCY

valve *n* a device (eg the triode) which was invented about the beginning of this century for amplifying weak electrical signals and which was subsequently used in early radio broadcasting. Valves have not been entirely replaced by semiconductors; they are still used for high-power switching and amplification. □ see TRIODE VALVE, THERMIONIC EMISSION, CATHODE-RAY TUBE

varactor diode *n* VARICAP DIODE

¹variable *n* a symbol (eg ×) in mathematics or a computer program which can be given a value in a specified range – compare CONSTANT

²variable *adj, of an electronic component* having a value which can be adjusted <*a ~ capacitor*>

varicap diode *n* also **varactor diode** a semiconductor diode specially made to have a relatively large capacitance between its cathode and anode terminals which can be varied by changing the reverse bias voltage across these terminals. Varicap diodes are used to tune in

stations in frequency modulation (FM) receiver circuits and TVs.
□ see AUTOMATIC FREQUENCY CONTROL

VCR – see VIDEO CASSETTE RECORDER

VDT *n* [*v*ideo *d*isplay *t*erminal] – see VDU

VDU *n* [*v*isual *d*isplay *u*nit] also **VDT** [*v*isual *d*isplay *t*erminal] an
input/output device having a screen and a keyboard to enable people
to communicate with a computer. Most VDUs use a cathode-ray
tube as the screen, although flat-screen displays, such as the liquid
crystal display, are becoming more common, esp in portable
computers. Keyboards, too, are becoming less useful and are being
replaced by the mouse, the touchscreen, and other input devices.

vector *adj* represented by direction as well as a size <*a line on a
VDU is a ~ quantity*> – compare SCALAR □ see VECTOR GRAPHICS

vector graphics *n* a method of reproducing fast-moving graphics on
a VDU by moving the electron beam in any direction across the
screen to leave a trace. Vector graphics uses a specially designed
cathode-ray tube which is smaller than the normal VDU.

Venn diagram *n* a diagrammatic method of identifying sets of data
which have common characteristics. A Venn diagram consists of a
square in which overlapping circles represent logic relationships
(AND, OR, etc). □ see KARNAUGH MAP

Venus Radar Mapper (abbr **VRM**) the name of an interplanetary
probe due for launch from Earth in 1987 and destined to reach
Venus planet in 1988. There it will be placed in orbit by remote
control from Earth. On-board instruments will use radar to
penetrate the cloud cover of Venus and map the whole Venusian
surface during the course of a year. The radar images will reveal
features as small as 1 kilometre across. The pictures sent back to
Earth will help astronomers understand the history of this planet.
□ see RADAR, INTERPLANETARY PROBE, VOYAGER, GALILEO

vertical metal-oxide semiconductor *n* VMOS

very high frequency *adj or n* VHF

very large scale integration *n* (abbr **VLSI**) the process or
technology of putting in excess of 5000 logic gates on a single silicon
chip. Microprocessors and some memory devices feature very large
scale integration. The trend is towards the integration of still more
transistors on a chip. The small size of individual components on
such chips is illustrated by a simple analogy: if a 5 millimetre square
chip of silicon is represented by a map of the UK, the smallest
component on it would occupy the area of a tennis court. □ see
MOORE'S LAW, PHOTOLITHOGRAPHY, FIFTH GENERATION COMPUTER

VHF *adj or n* [*v*ery *h*igh *f*requency] (of or being radio waves having)
a frequency in the range of 30 million hertz to 300 million hertz.
VHF is used mostly for high quality radio broadcasts (FM) and for
television broadcasts. □ see FREQUENCY MODULATION, UHF

VIA [*v*ersatile *i*nterface *a*dapter] a type of integrated circuit enabling
some microcomputers to handle a variety of signal conversions (eg

analogue-to-digital) between the computer and the outside world
□ see INTERFACE, PIA, USER PORT

¹video *n* VIDEO CASSETTE RECORDER

²video *adj* of television and the processing and/or displaying of images on a VDU <*a ~ amplifier*> □ see TELEVISION, VIDEO BUFFER, VIDEO AMPLIFIER, VIDEO CASSETTE RECORDER

video amplifier *n* **1** a device used to increase the strength of a television (video) signal produced by a television camera before being transmitted by radio **2** a device for increasing the strength of a television (video) signal in a television receiver before being delivered to a cathode-ray tube □ see TELEVISION, TELEVISION RECEIVER

video buffer *n* a part of a computer's random-access memory which holds the bit patterns of characters to be displayed on a VDU. This area of memory is periodically scanned and its readings passed to a character generator which forms the two-dimensional character on the screen. □ see BUFFER

video camera *n* a device for recording sound and vision on magnetic video tape for playback on a TV set. A video camera combines optics and electronics in one unit. A lens system focusses an image of the scene to be recorded on a light sensitive surface which is scanned to convert the image into a pattern of electrical (video) signals. These signals are added to the sound signals from a microphone and recorded on a video cassette. Video cameras are becoming smaller and lighter and their continued development, esp in the use of CCD imaging circuits, threatens the future of optical movie cameras. □ see VIDEO TAPE, VIDEO CASSETTE RECORDER, CHARGE-COUPLED DEVICE, TELEVISION

video cassette recorder *n* (abbr **VCR**) also **video** a device for recording television programmes and for playing back prerecorded programmes on magnetic video tape stored in video cassettes. A domestic VCR uses a helical scan system to lay down sloping video tracks on the magnetic tape. The tape is moved past a rapidly rotating head drum (read/write head) which magnetizes the coating on the tape with picture information. There are three main 'formats' for VCRs, and each of these is incompatible with any other. □ see VIDEO TAPE, VIDEO DISK, VIDEO CAMERA

videoconferencing *n* TELECONFERENCING

video disk *n* a disk on which is recorded, in digital form, picture and sound information for playing on a video disk player. Several different and incompatible methods have been developed for putting the information on the disk and for playing it back. Two well-developed systems are LaserVision and SelectaVision. Both systems store the picture information as a series of microscopic pits on a fine spiral track on the disk. The main difference between the two systems is the way that the information is picked up as the disk rotates at high speed in the player. A SelectaVision disk is made of conductive plastic so that the pits are seen by a special pickup head in contact with the surface of the disk as small changes in electrical capacitance. However, the system is very sensitive to the effects of dust on the surface of the disk and the freezing of single frames is difficult. LaserVision does not suffer from these problems. Its disk is also made of plastic but it is coated with a silvery metallized layer. The pickup head contains a tiny laser which directs a narrow beam of light at the pits on the track and a sensor picks up the variation in the light reflected from the pits and converts it into a stream of digital information for feeding to a VDU. The disk is covered with a layer of transparent plastic film so that any dust on the disk is out of focus to the sensor (just as dust on a camera lens does not appear as an image on the film). The pickup head is guided accurately over the track by information carried on the track along with the picture and sound information. The head can be moved to a single frame of a one hour programme. These disks enable information to be obtained quickly from any part of the program and are ideal for educational use since both text and pictures are stored and a programme can be made interactive. However, video disks do not allow a programme to be recorded from the TV in the home. □ see DIGITAL RECORDING, COMPACT DISK

video disk player *n* – see VIDEO DISK

video signal *n* the electrical signal produced in a television camera, or received in a television receiver, which is related to the variation in light (and colour) in the image being televized □ see TELEVISION, TELEVISION RECEIVER, VIDEO AMPLIFIER

videotape *vt* to record a TV programme, or to make a video film using a video camera, on video tape □ see VIDEO CASSETTE RECORDER, VIDEO DISK

video tape *n* a magnetic tape on which programmes are recorded in sound and vision, and which is held in a video cassette for playing on a video cassette recorder □ see VIDEO DISK

videotex *n* VIDEOTEXT

videotext *also* **videotex** *n* any system using VDUs to display computer-based information to people at home and work. There are two main types of videotext: broadcast videotext (known as teletext); and interactive videotext (known as viewdata). □ see VIEWDATA, TELETEXT

viewdata *n* also **interactive videotext** an information service which allows telephone subscribers access to a wide range of information from a database which is displayed on a TV set coupled to the telephone line by a modem □ see PRESTEL, PAGE, MODEM, MICRONET 800

viewdata editor *n* a software package which enables the user of a microcomputer to create pages of text and graphics for display on a VDU. The procedure is slow since the keys used are not dedicated to particular alpha-mosaic graphics symbols. However, the pages

produced can be shown one after the other on a carousel basis. The
method is particularly attractive to large companies which want to
keep their staff informed of company news, and to shopkeepers,
travel agents, etc.

Viking either of two spacecraft launched from the Earth in the 1970s
to land on Mars. The Viking spacecraft carried electronic equipment
to send back pictures from the surface of Mars and the results of
scientific investigations made by its small on-board laboratory. □ see
VOYAGER, GALILEO, SPACE SHUTTLE

virgin *adj, of a data storage medium* not having had data written into
it <~ *tape*>

virtual *adj, of an electronics system* appearing to have more facilities
than those actually provided by the hardware □ see VIRTUAL
MEMORY

virtual memory *also* **virtual storage** *n* a method used on large
computers of apparently providing extra memory which is not
restricted to the size of a computer's built-in memory. The
programmer identifies virtual memory locations which are translated
into actual locations using backing stores (eg magnetic tape) as
required.

virtual storage *n* VIRTUAL MEMORY

vision carrier *n* a radio wave which carries picture information (eg
television broadcasts on the UHF waveband) – compare SOUND
CARRIER □ see CARRIER WAVE

visual display terminal *n* VDU

visual display unit *n* VDU

VLSI – see VERY LARGE SCALE INTEGRATION

VMOS [*v*ertical *m*etal-*o*xide *s*emiconductor] a metal-oxide
semiconductor field-effect transistor (MOSFET) which is constructed
along the sides of a V-shaped groove which has been etched into the
silicon surface of a chip. This type of construction has the advantage
for integrated circuits that it requires a much smaller chip area than
for the planar type of construction. Discrete MOSFETs constructed
in this way are used for high power switching and amplifying
applications (eg audio amplifiers). □ see METAL-OXIDE
SEMICONDUCTOR FIELD-EFFECT TRANSISTOR

voice channel *n* a communications channel (eg the telephone line)
designed to carry human speech. The channel can carry sounds as
either analogue or digital signals. □ see TELEPHONE, SYSTEM X

voice input *also* **voice recognition** *n* the recognition of the human
voice by electronic circuits □ see SPEECH RECOGNITION, FIFTH
GENERATION COMPUTER

voice output *also* **voice response** *n* the production of speech by
artificial means (eg a computer) □ see SPEECH SYNTHESIS

voice recognition *n* VOICE INPUT

voice response *n* VOICE OUTPUT

volatile *adj, of a computer memory* needing constant electrical input

if stored data is not to be lost from it $<\sim RAM>$ – compare
NONVOLATILE □ see RANDOM-ACCESS MEMORY
volt n (symbol **V**) the unit of electrical potential difference; the
difference of electrical potential that will cause a current of 1 ampere
to flow through a resistance of 1 ohm □ see OHM'S LAW,
ELECTROMOTIVE FORCE
voltage n the number of volts at a point in a circuit, usu measured
relative to the circuit's zero potential □ see POTENTIAL DIFFERENCE,
VOLT
voltage comparator n – see COMPARATOR, SCHMITT TRIGGER
voltage divider n a circuit building block consisting of two or more
resistors in series which is connected across a power supply to
provide a useful lower voltage. Voltage dividers are used to provide
the correct operation of transistors in discrete and integrated circuits.
□ see POTENTIOMETER
voltage gain n – see GAIN
voltage level n the value of a voltage usu representing the high
(logic 1) and low (logic 0) levels needed to operate digital logic
circuits □ see THRESHOLD
voltage regulator n a three-terminal semiconductor device for
converting a varying DC voltage into a steady DC voltage. Voltage
regulators are very important components in many circuits, esp
digital circuits using TTL integrated circuits. □ see ZENER DIODE,
TRANSISTOR-TRANSISTOR LOGIC
voltmeter n an instrument for measuring electrical potential in volts
between two different points in a circuit □ see MULTIMETER, OHM'S
LAW, AMMETER
volume n the strength of an audio signal $<the \sim of\ sound>$
volume control n a control device which acts as a variable resistor
for altering the strength of an electrical signal $<a\ radio \sim>$ □ see
POTENTIOMETER
von Neumann, John. Contributed original ideas to mathematics
and in the Second World War was involved with ENIAC, the first
valve computer to be built in the USA. His work on computer design
laid the foundations for the modern electronic digital computer.
□ see VON NEUMANN MACHINE
von Neumann machine n one of the present generation of
computers based on concepts introduced by John von Neumann in
1946. In a von Neumann machine instructions are stored and
executed one after the other under the control of a central
processing unit. □ see VON NEUMANN, FIFTH GENERATION
COMPUTER
Voyager either of two spacecraft launched from the Earth in the late
1970s to examine at close range the planets Jupiter, Saturn, Uranus,
and Neptune. The Voyager spacecraft uses many on-board electronic
instruments to send back pictures and other data from interplanetary
space. Contact with the spacecraft is expected to continue until they

leave the solar system in the 1990s. □ see VIKING, GALILEO, SPACE SHUTTLE

VRM – see VENUS RADAR MAPPER

W

W the symbol for the unit of (electrical) power, the watt □ see WATT

wafer *n* also **slice** a thin disk cut from a single crystal of silicon on which hundreds of integrated circuits are made using the processes of photolithography and etching. The wafer is cut up into individual silicon chips which are then tested before being encapsulated in a plastic package. □ see BOULE, SILICON CHIP, PHOTOLITHOGRAPHY

walking aids *n* microelectronic aids to help disabled people to get about. The most exciting recent development helps people to walk who have lost the use of their legs through spinal injury. A computer is programmed to send electrical signals to the leg muscles which contract and relax in the correct sequence for walking. □ see BLIND AIDS, DEAF AIDS

Walter, Grey. A British neurologist and cybernetician who experimented with mechanical 'tortoises' in the early 1950s. Seymour Papert called the wheeled robots he operated under computer control 'turtles' in honour of Walter's work. □ see TURTLE, PAPERT, CYBERNETICS

wand *n* LIGHT PEN

warm boot *also* **warm start** *n* the way of getting a computer which is already switched on, ie one which is 'warm', into its correct operational status. Warm boot might be achieved using a 'reset' key, or by loading the disk operating system – compare COLD BOOT □ see OPERATING SYSTEM

warm start *n* WARM BOOT

watt *n* (symbol **W**) the unit of power, equal to a rate of conversion of energy of 1 joule per second. A light-emitting diode might convert 20 milliwatts (20mW) of electrical energy into heat and light.

wattage *n* the amount of power, measured in watts, generated or absorbed by a device. A soldering iron might have a wattage of 15.

wave *n* a regular variation in light, sound, voltage, pressure, or another property, which carries energy from one place to another <*radio* ~s> □ see ELECTROMAGNETIC WAVE, SINE WAVE, CARRIER WAVE

waveband *n* a range of radio or other frequencies <*the medium* ~> □ see BANDWIDTH, RADIO

waveform *n* the shape of a signal as displayed on a cathode-ray tube or plotter <*the* ~ *of an electrocardiogram*> □ see WAVEFORM GENERATOR

waveform generator *n* also **signal generator, function generator** an integrated circuit or a complete unit which produces electrical signals of useful shapes (eg sine, square, and triangular waveforms). A waveform generator is able to change the frequency of these waves from about 0.001 hertz to about 100 kilohertz, so it can be used for testing the behaviour of circuits (eg audio amplifiers) at different frequencies.

waveguide *n* a hollow metal tube along which radio waves carrying information travel without leaving its sides. The dimensions of the tube are usu equal to or greater than the wavelength of the microwaves which are usu in the range 2000 million hertz to 11 000 million hertz – compare OPTICAL FIBRE □ see MICROWAVES

wavelength *n* the distance between one point on a wave and the next corresponding point (eg from crest to crest). Wavelength is related to the frequency and speed of the wave by the simple equation: speed = frequency × wavelength. Thus a radio wave of wavelength 300 metres (medium wave) travelling through space at 300 million metres per second has a frequency of 1 million hertz (1 MHz). The tuning scale on some radios is sometimes marked with the frequencies of the stations rather than with their wavelengths. □ see FREQUENCY, PROPAGATION SPEED

weather satellite *n* a satellite put in orbit round the Earth to take measurements of atmospheric conditions to help in preparing weather forecasts. The information is transmitted to a receiving station as TV pictures and data. A weather station operates in polar orbit or in geostationary orbit. In polar orbit, the satellite passes over the North and South poles enabling a complete coverage of the weather patterns in the atmosphere to be obtained every 24 hours. In geostationary orbit, a continuous coverage is obtained of a single large area. One important sensing device aboard a weather satellite is a radiometer which scans the atmosphere to build up a picture of cloud cover and temperature at different altitudes, including the temperature of the sea. By measuring radiation emitted by the atmosphere in the infrared region, cloud pictures at night can be obtained. Information from weather satellites is available not only to weather forecasters but also to ships at sea, aircraft in flight, and to radio amateurs. □ see COMMUNICATIONS SATELLITE, GEOSTATIONARY

wetware *n* the human brain □ see LIVEWARE, FIRMWARE, SOFTWARE, HARDWARE

Wheatstone bridge *n* a device consisting of four resistors connected in series, used to provide a resistance-to-voltage conversion in the design of some electronic instruments. A power supply is connected across two opposite corners of the bridge and an output is obtained across the other two corners. One resistor is usu a transducer, for example a thermistor for measuring temperature change, a light-dependent resistor for measuring light change, or a strain gauge for measuring mechanical distortion. A small change in the value of transducer causes a small change in the output voltage which can then be measured. □ see TRANSDUCER

white noise *n* noise that sounds like background hiss and has no detectable frequency characteristic. On communications channels, white noise is caused by the random movement of electrons in wires and components. But white noise is put to good use in computer games by 'shaping' it into the noise of explosions or similar sound effects. Some instruments designed to relax the body use white noise to produce a soft 'swishing' sound like the noise of waves on a seashore. □ see NOISE

wideband *n* also **broadband** a range of radio frequencies wide enough to be divided into a number of narrower bands (eg for radio control). Generally wideband means a high capacity communications channel. □ see BANDWIDTH

Wien bridge oscillator *n* a circuit that uses resistors, capacitors, and an operational amplifier to generate sine waves used for testing and calibration of equipment □ see WAVEFORM GENERATOR

Winchester disk *n* a data recording and playback device in the form of a hard and rigid disk, about 200 millimetres (8 inches) or 130 millimetres (5¼ inches) in diameter, coated with magnetic material for storing data. The disk is driven at high speed in a chamber sealed against dust and dirt. A read/write head is used to transfer data to and from the disk and this 'floats' just above the surface of the disk on a cushion of air. □ see FLOPPY-DISK, FERRITE, READ/WRITE HEAD

window *n* **1** any of a number of 'openings' in the Earth's atmosphere through which light and radio waves from outer space can penetrate and reach the ground *<a ~ for radio astronomy>* **2** an interval of time within which an event must occur if it is to have a successful outcome *<the launch ~ for a flight of the space shuttle>* **3** a small part of the display on a VDU selected for special effects *<to enlarge a ~>*

wiper *n* the terminal of a variable resistor or potentiometer which makes contact with a conducting track so that a variable resistance can be selected □ see VOLUME CONTROL, POTENTIOMETER

¹wire *vb* to make an electrical connection between two or more components or between two circuits using electrically conducting wire

²wire *n* a single metal strand or collection of strands along which electricity flows easily □ see CABLE

wired *adj* connected together by means of wires *<a ~ logic circuit>*

wireless *n* the original (British) name for a radio receiver since, unlike the telephone, it worked without wires □ see RADIO

wirewrapping *n* the process of making electrical connections to integrated circuits and other components by binding silver-plated wire tightly round their terminal pins using a special tool. Good electrical connection is obtained by stretching the wire and ensuring that a terminal wire has at least two sharp edges along its length. – **wirewrap** *vb* □ see WIREWRAPPING SOCKET

wirewrapping socket *n* a socket for holding integrated circuits which has extra long terminal pins so that wire connections to them can be made using a special tool

wiring diagram *n* a drawing showing the connections between

components in an electrical circuit <*the ~ of a radio*> □ see BLOCK
DIAGRAM, SYSTEM

woofer *n* a loudspeaker (larger than a tweeter) used in high quality
audio systems for radiating low frequency sound – compare
TWEETER A □ see LOUDSPEAKER

word *n* a pattern of bits which is handled as a single unit of data by a
computer or other digital processing device <*an 8-bit ~*> □ see
NIBBLE, BYTE, BIT

word length *n* the number of bits (ie 0s and 1s) in a word □ see
WORD

wordprocessing *n* a computer-based method for producing,
editing, storing, printing, and transmitting typed information □ see
WORDPROCESSOR, ELECTRONIC OFFICE

wordprocessor *n* (abbr **WP**) a computerized typewriter which
allows written material to be generated, stored, edited, printed, and
transmitted. A wordprocessor has five main parts: a keyboard for
inputting text; a microprocessor for making decisions about editing
procedures; a VDU for displaying text being edited and manipulated
by the keyboard; a floppy disk drive or magnetic bubble memory
with which files can be accessed and stored; and a printer for
producing hard copy of the text. A wordprocessor may be bought as
a dedicated device for use in an office, or a microcomputer can be
turned into a wordprocessor using a software package on disk, tape,
or ROM. □ see TEXT EDITOR, ELECTRONIC OFFICE

working area *n* WORKING SPACE

working space *also* **working area** *n* any area of memory on disk or
tape available to a program – compare WORKSTATION

working voltage *n* the maximum voltage which a component, esp a
capacitor, can withstand across its terminals without fear of damage

workstation *n* a self-contained unit comprising a desk, chair, and
storage facilities for computing activities. A workstation might be
equipped with a VDU and keyboard linked to printers and modems,
as in an office, or it might be a simple affair for supporting a home
computer and a few peripherals, as in the home. □ see TERMINAL

wow *n* – see WOW AND FLUTTER

wow and flutter *n* unwanted 'wobble' to the pitch of sounds
produced by a tape recorder or record player if the tape or record is
not moving at a constant speed. This variation in pitch is called wow
if it is low-frequency, and flutter if the variation is at a higher
frequency. Compact disk record players do not suffer from wow and
flutter. □ see COMPACT DISK

WP – see WORDPROCESSOR

wrap *n* WRAPAROUND

wraparound *also* **wrap** *n* the process whereby a computer
automatically moves the cursor on the VDU to a new line when the
text has reached the right-hand side of the screen. A wordprocessor
uses word wraparound by which a whole word is moved to the next

line to avoid broken words.

wristwatch computer *n* a digital wristwatch providing computer functions as well as timekeeping, alarm, and calendar functions. One type of wristwatch computer displays 4 lines of program, each of 10 characters, on the watch's liquid crystal display. It has 2K of RAM, 6K of ROM, and four cursor and control keys. A separate 62-key keyboard sends information to the watch by means of radio waves. A printer and other peripherals are available for the watch. □ see DIGITAL WATCH, PORTABLE MICRO

wristwatch TV *n* a television receiver built into a wristwatch. At the present time, a separate battery pack has to be carried on the belt or in the pocket to power a wristwatch TV, but the pace of change in microelectronics is so fast that wristwatch TVs may soon be powered by a built-in power supply. It is very likely that they will enable us to see and speak to other people anywhere in the world, the signals being relayed by artificial satellites. □ see FLAT-SCREEN TV, LIQUID CRYSTAL DISPLAY

write *vb* to put data into a computer memory – compare READ □ see READ/WRITE HEAD

write enable *n* a signal which must be sent from a computer to enable a disk or tape to store data

write head *n* a magnetic device in tape and disk drives which puts data (eg programs) on magnetic tape or disk. The write head is usu part of the reading head – compare READ HEAD

X

X-band *n* a range of radio waves having frequencies from 5.2 gigahertz to 10.9 gigahertz used esp for high-frequency radar and communications with orbiting satellites and interplanetary probes □ see GIGAHERTZ

x coordinate *also* **x position** *n* a number in a graphics program which lets the computer know the horizontal position of a particular pixel it is to access on the screen of a VDU – compare Y COORDINATE

XNOR gate *n* EXCLUSIVE NOR GATE

XOR gate *n* EXCLUSIVE OR GATE

x position *n* X COORDINATE

X-ray lithography *n* the use of X-rays and an X-ray mask to define circuit detail on a silicon chip. X-ray lithography is the same, in principle, as the well-established technique of photolithography and, because of the much shorter wavelength of X-rays compared with ultraviolet light, it provides much finer definition (less than 1 micron) for components on the chip. However, X-ray lithography is proving to be a difficult technique to implement and it is not as well advanced as electron-beam lithography and photolithography. □ see X-RAYS, PHOTOLITHOGRAPHY, PHOTOMASK, ELECTRON-BEAM LITHOGRAPHY

X-rays *n* penetrating electromagnetic radiation used in industry and in medicine for 'seeing' below the surface of solid materials. X-rays have wavelengths much shorter than those of visible light and are being considered for use in the manufacture of integrated circuits. □ see X-RAY LITHOGRAPHY, ELECTROMAGNETIC WAVE

xtal *n* [crystal] – see CRYSTAL CLOCK

xy plotter *n* – see GRAPH PLOTTER

Y

Yagi aerial *n* DIPOLE AERIAL

y coordinate *also* **y position** *n* a number in a graphics program which lets a computer know the vertical position of a particular pixel it is to access on the screen of a VDU – compare X COORDINATE

yield *n* the percentage of good silicon chips manufactured in a single production run. A yield of 100 per cent is never achieved since the manufacture of silicon chips is a complex process and depends on many steps which have to be performed with great precision. The yield of a particular integrated circuit varies from about 99 per cent for chips containing a few thousand transistors to less than 1 per cent for complex very large scale integrated circuits such as microprocessors. □ see PHOTOLITHOGRAPHY, SILICON CHIP, VERY LARGE SCALE INTEGRATION

y position *n* Y COORDINATE

Z

Z the symbol for impedance □ see IMPEDANCE

zap *vb* to make small changes to data in a computer memory by unconventional methods which usu threaten the loss of all the data

Zener diode *n* also **breakdown diode** a special semiconductor diode which is designed to conduct current in the reverse direction (ie from its cathode to its anode) when a voltage above a certain value is applied in that direction. The Zener diode is used in power supply circuits to convert a varying voltage into a steady one, for suppression of fast-changing signals (eg transients), and for clipping off the top of waveforms. □ see VOLTAGE REGULATOR, CLIPPING, TRANSIENT, DIODE

¹zero *n* the binary or decimal digit which represents a quantity of zero numerical value

²zero *adj* having no value or size <~ *insertion force socket*>

³zero *vb* to adjust some part of hardware or software to a known starting condition

zero voltage switch *n* an integrated circuit designed for triggering a triac used for the control of alternating current power. The zero voltage switch minimizes the generation of radio frequency interference which triacs and similar power control devices generate when they switch on and off. The switch only triggers the triac at the point when the alternating current is zero. □ see TRIAC, SILICON-CONTROLLED RECTIFIER

zif socket *n* [*zero insertion force* socket] a holder which is soldered to a printed circuit board and in which a dual-in-line integrated circuit (IC) is held by a lever mechanism built into it. Use of the zif socket, rather than a conventional IC holder which has spring clips, ensures that the IC is gripped securely without applying too much pressure to the pins. □ see DUAL-IN-LINE

zinc air cell *n* a primary cell (ie nonrechargeable) which produces a stable output voltage, has a fairly long shelf life, and is a compact source of electrical energy. The zinc air cell is a relatively new type of cell which has an operating voltage of 1.4 volts and which is therefore more flexible to use than the lithium cell in electronic systems (eg hearing aids) – compare LITHIUM CELL

zone refining *n* a method of obtaining very pure silicon for making integrated circuits. The process involves passing a heated silicon rod slowly through a furnace so that a molten zone of silicon is made to move along the rod. By repeating the process several times, any impurities in the silicon are moved along to one end of the rod leaving most of the silicon very pure. □ see BOULE, PHOTOLITHOGRAPHY

zoom *vb* to enlarge the size of all or part of a computer graphics display □ see WINDOW

Zuse, Konrad. German designer and builder of a programmable calculator in the 1940s. His first device was mechanical and could perform basic arithmetic operations. Later he designed electrical calculators, one of which was used to help in the construction of the V1 and V2 flying bombs.